Journees Nationales des Spectroscopies de PhotoEmission (JNSPE 2022)

EPJ Web of Conferences Volume 273 (2022)

Dijon, France
16 – 18 May 2022

Editors:

N. Barrett
H. Martinez
A. Tejeda

ISBN: 978-1-7138-6621-3

Printed from e-media with permission by:

Curran Associates, Inc.
57 Morehouse Lane
Red Hook, NY 12571

CURRAN ASSOCIATES INC. proceedings.com

Some format issues inherent in the e-media version may also appear in this print version.

For additional information, please contact EDP Sciences – Web of Conferences
at the address below.

EDP Sciences – Web of Conferences
17, Avenue du Hoggar
Parc d'Activité de Courtabœuf
BP 112
F-91944 Les Ulis Cedex A
France

Phone: +33 (0) 1 69 18 75 75
Fax: +33 (0) 1 69 28 84 91

contact-edps@webofconferences.org

Additional copies of this publication are available from:

Curran Associates, Inc.
57 Morehouse Lane
Red Hook, NY 12571 USA
Phone: 845-758-0400
Fax: 845-758-2633
Email: curran@proceedings.com
Web: www.proceedings.com

TABLE OF CONTENTS

Preface:

The JNSPE2022 conference was the first national event of the CNRS (National Center for Scientific Research) "Photoemission Spectroscopies" Research Federation (FR SPE N° 2050 - https://fr-spe.cnrs). The objective of the Federation is to structure the scientific community around photoemission and related spectroscopies by sharing the diverse expertise present in laboratories on the national territory. This research structure project in the various fields covered by the Photoemission Spectroscopies is the result of a broad upstream reflection initiated by the CNRS Institute of Chemistry (INC@CNRS) in 2019 and was officially created in 2020. Since then, the Federation organizes and coordinates visible and ambitious actions, with the constant concern to promote interdisciplinarity among its staff belonging to 48 CNRS laboratories. They are part of the Institutes of Chemistry (INC), Physics (INP) and Engineering and Systems Sciences (INSIS) as well as a CEA (French Alternative Energies and Atomic Energy Commission) laboratory (PFNC) and IFPEN (IFP Innovative Energies). A total of 340 researchers, university professors, engineers and technicians from a multitude of backgrounds are part of this open structure. The Federation works on identified scientific challenges, creating very strong thematic synergies between teams and even laboratories. It also invests in training activities adapted to the different needs of the scientific community. It is an ideal place to reflect on the structuring of the laboratory spectrometer network community, as well as on the establishment of open platforms.

The JNSPE2022 conference drew on the achievements of the two national congresses ELSPEC (2004-2018) and JSE (2017-2021) to bring together the national and French-speaking community of specialists and users of photoemission spectroscopy in the broad sense around a single annual event. This national event is also an opportunity to open up to researchers and engineers who may envisage using photoemission techniques.

The aim of the meeting is to present the latest developments in Photoemission Spectroscopies, both for the understanding of phenomena related to the physics and chemistry of surfaces and interfaces and for the analytical aspect in the broad sense of these techniques. The French-speaking community of photoemission spectroscopies is in fact largely involved in many themes, such as (non-exhaustive list): energy storage and conversion, micro- and nanotechnologies, metal alloys and corrosion, functional oxides, catalysis, electrochemistry, nanomaterials, thin layers and growth phenomena, biomaterials and interface engineering, surface functionalization, synthesis of organic hybrids...

These themes constitute the core skill of the 48 research CNRS laboratories of the federation and will be addressed in the light of the scientific challenges identified by the spectroscopic community for the next five years. Over five half-days (June 16-18, 2022) punctuated by invited conferences of general interest, participants of the JNSPE 2022 (Dijon) shared their latest research on:

- The environmental XPS approach
- The problem of buried interfaces
- Aspects of operando experiences
- Measurement reliability and quantification
- Photoemission in its time dimension
- Experimental and theoretical approaches in photoemission for the study of electronic structures

120 participants and 8 specialist equipment manufacturers discussed the latest advances in the field over 2.5 days via 41 oral communications, 25 posters. More, five internationally recognized guest speakers were able to introduce the different sessions and provide relevant insights on the major issues facing our scientific community:
Virginia PÉREZ-DIESTE (Synchrotron ALBA, Spain): Operando characterization of model and real catalysts by APXPS.
Thierry CONARD (IMEC, Belgium): Photoemission in microelectronics: When technological developments allow to address new questions in the laboratory.
Mario EL KAZZI (PSI, Switzerland): $LiCoO_2$ interface electrochemical and electronic properties – Li_3PS_4 reviewed by operando XPS in all-solid battery.
Mathieu G. SILLY (Synchrotron Soleil, Saint-Aubin): Study of the relaxation dynamics of photoexited loads by time-resolved photoemission spectroscopy.
Marc SIMON (LCPMR, Paris): Gas phase photoemission study.

The present conference proceedings provide a representative overview of the diversity of the scientific contributions as well as papers from each of the guest speakers for this first edition of the JNSPE.

Hervé Martinez
Director of the FR SPE

edp sciences

EPJ Web of Conferences

Statement of Peer review

In submitting conference proceedings to EPJ Web of Conferences, I certify to the Publisher that I adhere to the **Policy on Publishing Integrity** of the journal in order to safeguard good scientific practice in publishing.

1. All articles have been subjected to peer review administered by the proceedings editors.
2. Reviews have been conducted by expert referees, who have been requested to provide unbiased and constructive comments aimed, whenever possible, at improving the work.
3. Proceedings editors have taken all reasonable steps to ensure the quality of the materials they publish and their decision to accept or reject a paper for publication has been based only on the merits of the work and the relevance to the journal.

Title, date and place of the conference

JNSPE 2022 16-18 Mai 2022 Dijon

Proceedings editor(s):

Nick BARRETT
Hervé MARTINEZ
Antonio TEJEDA

Date and editor's signature

5 octobre 2022

17, avenue du Hoggar - PA de Courtabœuf – BP 112 - 91944 Les Ulis Cedex A (France)
Tél. : 33 (0)1 69 18 75 75 - Fax : 33(0)1 69 07 45 17 – www.edpsciences.org

Li-ion solvation in TFSI and FSI -based ionic liquid electrolytes probed by X-ray photoelectron spectroscopy

Mario El Kazzi [1*]

[1] Electrochemistry Laboratory, Paul Scherrer Institute, 5232 Villigen PSI, Switzerland

Abstract. For Li-ion batteries, the Li-ion solvation in liquid electrolytes is a crucial parameter affecting directly the electrochemical cycling performance. X-ray photoelectron spectroscopy (XPS) can play an essential role for investigating the cation and anion electronic structure and monitoring the Li-ion solvation into various solvent and salt environments. In this contribution, we demonstrate the capability of conventional laboratory XPS using Al Kα X-ray source to determine the anions solvation shell of Li^+ cation within the low vapour pressure and vacuum compatible ionic liquid electrolytes. 1M of LiTFSI and 1M of LiFSI salts dissolved in ($EMIM^+$-FSI^-) and ($EMIM^+$-$TFSI^-$) ionic liquids respectively are investigated by acquiring the F1s, N1s, C1s, S2p and Li1s core levels. The binding energy difference between the N1s component originating from the $EMIM^+$ cation and the N1s component originating from $TFSI^-$ or FSI^- anions solvating the Li^+ confirms that both $TFSI^-$ and FSI^- contribute simultaneously to the Li^+ solvation. Additionally, the stability of the TFSI and FSI -based ionic liquid electrolytes is carefully discussed for long X-ray exposure times.

1 Introduction

The continuous need for increasing the energy density, safety and electrochemical performance of Li-ion batteries (LiBs) requires an incessant development of characterization techniques capable of providing better insights into the physics and chemistry of the various parts in a battery.[1] Upon cycling, both the bulk and the surface of the electrode's components are modified with a direct impact on the battery performance. When explicitly investigating the surface and interface modifications, X-ray photoelectron spectroscopy (XPS) is routinely employed and considered to be one of the suitable characterization techniques for LiBs allowing to study the interface evolution occurring on cycled electrodes. [2, 3] The advantage of the XPS is the surface sensitivity providing direct information of the by-product species related to the electrolyte stability, in both liquid [4] and solid [5],[6],[7] -based electrolytes. Furthermore, by probing the active materials, XPS offers the possibility to identify the potential dependency of their oxidation states and redox reactions. Moreover, by using synchrotron facilities and tuning the X-ray energy source from soft to hard, a non-destructive depth profile analysis can be performed by probing deeper from the surface down to 50 nm.[8], [9] The usage of XPS in LiBs goes beyond the *ex-situ* investigation of cycled electrodes. *Operando* XPS in all-solid-state batteries can probe the electrified solid/solid interface between the active materials and solid electrolyte. This was recently demonstrated to provide valuable information concerning the (electro-)chemical and electronic properties of the interface.[10],[11] In addition, probing the electrified liquid/solid [12] interface have been attempted, for instance between the ionic liquid electrolyte and the Li metal anode.[13, 14]
XPS in LiBs is not limited to study the electrolyte/electrode interface, it can also be carried out to probe just the liquid electrolyte and investigate the Li-ion solvation for example in carbonate -based electrolytes as already demonstrated by our group using near ambient pressure photoemission (NAP) combined with a liquid jet.[15] Such experiment and information are very precious, as the Li-ion solvation in the LiBs' electrolyte play an important role in improving the cycling performance, by directly affecting the ionic conductivity, the stability at high or low potential and temperature, as well as the property of the solid electrolyte interphase (SEI).
In this work, conventional XPS using Al Kα X-ray source is employed to investigate the cation and anion electronic structure and to monitor the Li-ion solvation within droplets of low vapour pressure and vacuum compatible ionic liquid electrolytes.
Ionic liquid-based electrolytes are considered as promising alternative for next generation high energy density LiBs owing to their low vapour pressure and high thermal as well as electrochemical stability. However, not all ionic liquid-based electrolytes can successfully be deployed in LiBs. For example, stable electrochemical cycling of graphite in ionic liquid electrolytes can be achieved in presence of FSI^- anion originating either from the LiFSI salt or as an anion from the ionic liquid.[16] On the contrary, graphite cannot be cycled when $TFSI^-$ anion is used, unless FSI^- is added in the electrolyte, despite the structural similarity between

* Corresponding author: mario.el-kazzi@psi.ch

the anions. Intrigued about this phenomenon, we examined in this study the Li-ion solvation when both $TFSI^-$ and FSI^- are present simultaneously in the electrolyte, as a salt or as anion in the ionic liquid. 1M of LiTFSI and 1M of LiFSI salts dissolved in ($EMIM^+$ - FSI^-) and ($EMIM^+$ - $TFSI^-$) ionic liquids respectively are investigated by acquiring the F1s, N1s, C1s, S2p and Li1s core levels. The abbreviation of EMIM, TFSI and FSI stand for 1-ethyl-3-methylimidazolium, bis(trifluormethanesulfonyl) imide and bis(fluorosulfonyl) imide respectively (Fig.1).

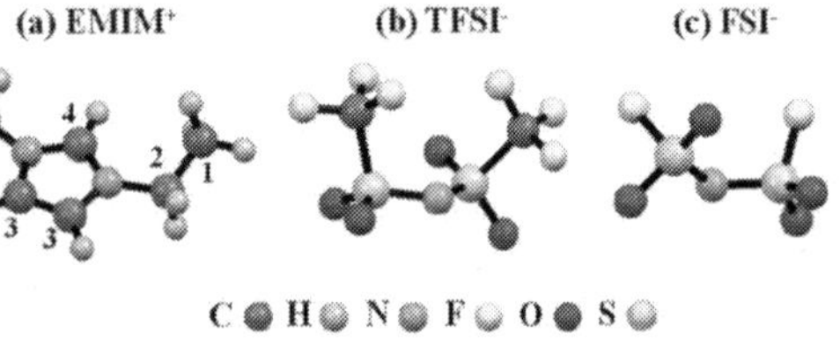

Fig. 1. Structural formula of (a) $EMIM^+$ cation, (b) $TFSI^-$ and (c) FSI^- anions.

2 Experiments

XPS measurements were performed with a VG ESCALAB 220iXL spectrometer (Thermo Fisher Scientific) with a base pressure of $\sim 2 \times 10^{-9}$ mbar using a focused monochromatized Al Kα radiation (1486.6 eV) with a beam size of $\sim 500\ \mu m^2$. The spectrometer was pre-calibrated by performing a measurement on a clean silver surface, whereby the Ag $3d_{5/2}$ peak was aligned to a binding energy of 368.25 eV with a full width at half-maximum of 0.78 eV at a pass energy of 30 eV. All the narrow energy span spectra are recorded with a dwell time of 50 ms, using a pass energy of 30 eV in steps of 0.05 eV. Peak deconvolution is performed by applying the sum of Gaussian (70%) and Lorentzian (30%) line shapes after a Shirley-type background subtraction. The peaks binding energy alignment is done on the N1s originating from the $EMIM^+$ cation component located at 402 eV.

All the ionic liquids and electrolytes are dried at 80 °C under 10^{-2} mbar in a schlenk line for more than 6 hours and transferred under vacuum to the Ar filled glovebox. The XPS measurements are performed on droplets deposited on the electronically conductive active carbon substrate transferred from the glovebox to the XPS load lock using a dedicated transfer chamber sealed under Ar atmosphere. First, the XPS measurements are performed on two reference samples of pure [$EMIM^+$-$TFSI^-$] and [$EMIM^+$-FSI^-] ionic liquids and then on four ionic liquid electrolytes where 1M of LiTFSI and 1M of LiFSI salts are dissolved in either [$EMIM^+$ - FSI^-] or [$EMIM^+$ - $TFSI^-$] ionic liquids.

3 Results

3.1 [$EMIM^+$-$TFSI^-$] and [$EMIM^+$-FSI^-] ionic liquids

The XPS core levels acquired on the [$EMIM^+$-$TFSI^-$] and [$EMIM^+$-FSI^-] ionic liquids presented in Fig. 2 are used as reference spectra to identify the exact binding energy (BE) position and full width half maximum (FWHM) of the peaks originated from both cation and anion. The extracted parameters BE and FWHM will be used to deconvolute precisely the components originating from the Li^+ cation and its solvated shell in the electrolytes.

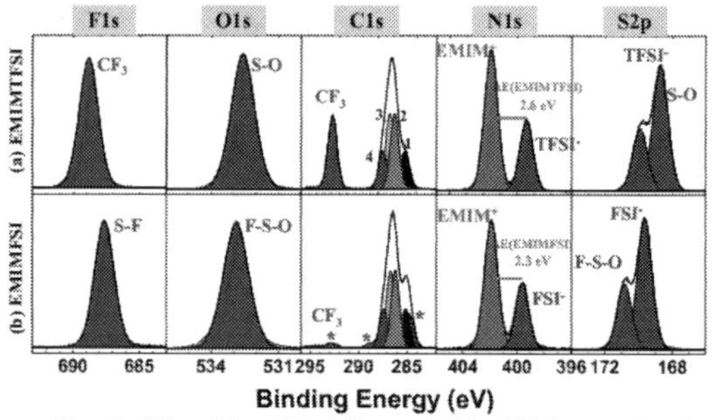

Fig. 2. F1s, O1s, C1s N1s and S2p XPS core levels acquired on (a) [$EMIM^+$-$TFSI^-$] and (b) [$EMIM^+$-FSI^-] ionic liquid droplets.

In the [$EMIM^+$-$TFSI^-$] solution (Fig.2.a) we distinctly identify in the N1s core level the components located at 402 eV and 399.4 eV associated with the N atom in the $EMIM^+$ and $TFSI^-$ respectively with a BE difference $\Delta E = N1s_{(EMIM)} - N1s_{(TFSI)} = 2.6$ eV. The C1s core level is comprehensibly deconvoluted with the components originating from $EMIM^+$ (Fig.1.a) related to the (1) aliphatic carbon CH_x, (2) C-N bonds, (3) C-C*-N and (4) N-C*-N located at 285.2 eV, 286.3 eV, 286.8 eV and 287.5 eV respectively. The CF_3 component located at 292.7 eV, which originates from the $TFSI^-$ (Fig.1.b). As expected both F1s and O1s show a single component located at 688.8 eV and 532.6 eV respectively, while the S2p is deconvoluted with the j=3/2 and j=1/2 spin-orbitals located at 168.7 eV and 169.9 eV respectively. The F1s, O1s and S2p signals originate all from the $TFSI^-$ (Fig.1.b). No traces of contamination or ionic liquid degradation under the X-ray source is detected in the solution.

In the [$EMIM^+$-FSI^-] solution (Fig.2.b) we can still identity with the same level of distinction both the N1s components located at 402 eV and 399.7 eV from $EMIM^+$ and FSI^- respectively but with reduced BE difference $\Delta E = N1s_{(EMIM)} - N1s_{(FSI)} = 2.3$ eV. The C1s components associated with the $EMIM^+$ are deconvoluted in the same way as above. However, additional non expected components located at *284.9 eV, *289 eV and *292.7 eV had to be added in order to satisfy the fit. Those additional components are affiliated with C-H, O=C-O and CF_3 impurities or EMIMFSI degradation. We cannot also exclude signals coming from the active carbon even though the substrate is soaked by the ionic liquid and forming a large droplet on the surface. The F1s, O1s and S2p still detected as single peaks located at 687.6 eV, 532.9 eV and 169.6 eV respectively, all originating from FSI^- (Fig.1.c). The atomic percentage of the different elements present in EMIMTFSI and EMIMFSI electrolyte are displayed in Table 1. An error of 2% is expected calculated on a clean Cu foil reference using the Cu2p, Cu3s and Cu3p core levels.

Table 1. Estimated atomic percentage (%) of on the elements present in [$EMIM^+$-$TFSI^-$] and [$EMIM^+$-FSI^-] ionic liquids.

$N1s_{cation}$	$N1s_{anion}$	O1s	S2p	F1s	C1s
EMIMTFSI					
11%	6%	21%	11%	39%	12%
EMIMFSI					
16%	9%	35%	19%	21%	-

3.2 Li^+ solvation by $TFSI^-$ and FSI^- anions

By dissolving 1M of LiTFSI in [$EMIM^+$-$TFSI^-$] we can be sure that the Li^+ cation is solely solvated by $TFSI^-$ anions ($TFSI^-$–Li^+–$TFSI^-$). Similarly, by dissolving 1 M of LiFSI in [$EMIM^+$-FSI^-], Li^+ cation is solely solvated by FSI^- anions (FSI^-–Li^+–FSI^-). The N1s and S2p core levels acquired on those electrolytes are presented in Fig.3.a and b. Additional components are detected as soon as the salts are added. In the N1s peaks the extra components are located at 399.85 eV/400.1 eV between the $EMIM^+$ and $TFSI^-$/FSI^- components while in the S2p it is located at high BE from $TFSI^-$/FSI^- components.

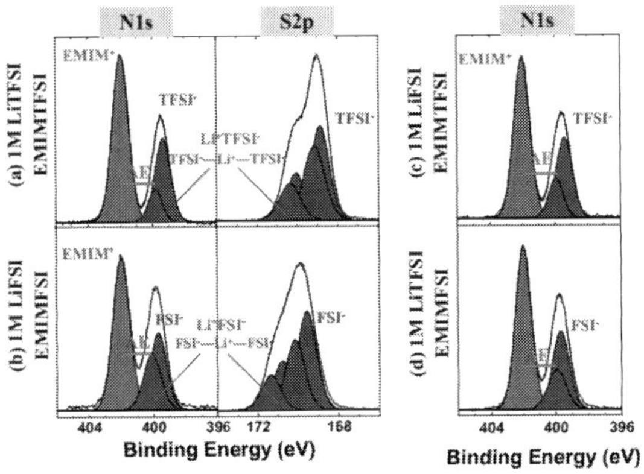

Fig. 3. N1s and S2p XPS core levels acquired on (a) 1M LiTFSI in [$EMIM^+$-$TFSI^-$] and (b) 1M LiFSI in [$EMIM^+$-FSI^-]. N1s acquired on (c) 1M LiFSI in [$EMIM^+$-$TFSI^-$] and (d) 1M LiTFSI in [$EMIM^+$-FSI^-].

The signal of those additional components originate from the $TFSI^-$/FSI^- anions solvating the Li^+ cations called hereafter Li^+TFSI^- and Li^+FSI^-. The BE difference between the $EMIM^+$ and the Li^+TFSI^- [$\Delta E = N1s_{(EMIM)} - N1s_{(LiTFSI)}$] or Li^+FSI^- [$\Delta E = N1s_{(EMIM)} - N1s_{(LiFSI)}$] components are calculated to be 2.15 eV and 1.9 eV respectively. As the $EMIM^+$ cation is present in all the investigated electrolytes, so it is used as reference unchangeable component. Based on the assumption that its electronic structure will not change, as it is not expected to participate in the first solvation shell of the Li^+. Thus, the measured ΔE can be assumed as characteristic parameter linked directly to the Li^+ solvation in the electrolyte. To summarise, when Li^+ is fully solvated by $TFSI^-$ the resulting ΔE is equal to 2.15 eV while when Li^+ is fully solvated by FSI^- ΔE is reduced to 1.9 eV.

By applying the same methodology, we are interested in understanding the mechanism of the Li^+ solvation when both $TFSI^-$ and FSI^- are present simultaneously in the electrolyte as salt or as anion in the ionic liquid. For that, the N1s core level acquired on both 1M LiFSI in [$EMIM^+$-$TFSI^-$] and 1M LiTFSI in [$EMIM^+$-FSI^-] electrolytes are presented in Fig.3.c and d. The BE difference $\Delta E = N1s_{(EMIM)} - N1s_{(solvated\ Li)}$ is calculated to be 2.07 eV and 2.06 eV for 1M LiFSI in [$EMIM^+$-$TFSI^-$] and 1M LiTFSI in [$EMIM^+$-FSI^-] respectively, which is between the ΔE measured above when Li^+ is fully solvated by $TFSI^-$ and FSI^-. This result demonstrate that Li^+ is solvated by both $TFSI^-$ and FSI^- when both anions are present in the electrolyte as salt or in the ionic liquid. All the calculated ΔE are reported the table 2 below.

The same trend is obtained when the BE energy difference is calculated between [$\Delta E = N1s_{(EMIM)} - S2p_{(solvated\ Li)}$] or [$\Delta E = N1s_{(solvated\ Li)} - Li1s_{(solvated\ Li)}$]. However, we do not recommend using the Li1s core level with the risk to introduce error on the measurements caused by the electrolytes instability under the X-ray source and the formation of LiF during the acquisition as discussed in the next paragraph.

Table 2. Calculated $\Delta E = N1s_{(EMIM)} - N1s_{(solvated\ Li)}$ for the different prepared ionic liquid electrolytes

N1s	ΔE=N1s(EMIM)-N1s(solvated Li)	
1M LiTFSI $EMIM^-$ $TFSI^+$	2.15 eV	$TFSI^-$…Li^+…$TFSI^-$
1M LiFSI $EMIM^-$ FSI^+	1.9 eV	FSI^-…Li^+…FSI^-
1M LiFSI $EMIM^-$ $TFSI^+$	2.07 eV	$TFSI^-$…Li^+…FSI^-
1M LiTFSI $EMIM^-$ FSI^+	2.065 eV	FSI^-…Li^+…$TFSI^-$

3.3 Electrolyte stability under the X-ray beam

The stability of the electrolytes under the X-ray source is investigated by acquiring the F1s core level. It is worth mentioning that the measured pure ionic liquids $EMIM^+$ $TFSI^-$ or $EMIM^+$ FSI^- without the salts show good stability under X-ray. No sign of any ionic liquid degradation is detected and no change in the XPS core levels is observed even after long exposure time beyond 2 hours. However, as soon as the salts are added in the ionic liquid, then LiF component is detected in the F1s core level located at 685 eV (Fig.4.). It is a chemical reaction between the Li-ion and the F from either TFSI or FSI. The amount of LiF increases with the exposure time. Some exception is observed for the 1M LiTFSI in [$EMIM^+$-FSI^-] (Fig.4.c), where the amount of LiF was high initially directly after exposure to X-ray and did not evolve after 2 hours. For the 1M LiFSI in [$EMIM^+$-FSI^-] an additional, unknown component is detected in the

F1s (fig.4.d) located at 687 eV besides the LiF, which both evolve after 2 hours exposure time to X-ray.

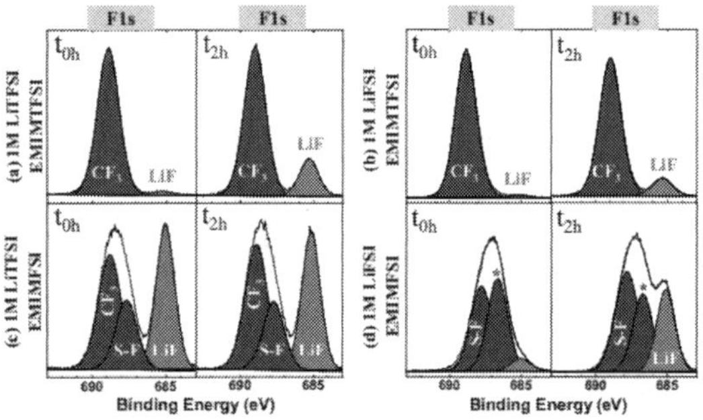

Fig. 4. F1s XPS core levels acquired directly after transfer (T_{0h}) and after 2 hours (T_{2h}) exposure to X-ray Al Kα for the electrolytes: (a) 1M LiTFSI in [$EMIM^+$-$TFSI^-$], (b) 1M LiFSI in [$EMIM^+$-$TFSI^-$], (c) 1M LiTFSI in [$EMIM^+$-FSI^-] and (d) 1M LiFSI in [$EMIM^+$-FSI^-].

4 Conclusion

In our study, we show a new approach for investigating the Li-ion solvation in TFSI and FSI -based ionic liquid electrolytes carried out by conventional laboratory XPS using Al Kα X-ray source. This is conceivable thanks to the low vapour pressure of the ionic liquid and its stability under ultra-high vacuum. We demonstrate that the Li^+ cation is solvated by both $TFSI^-$ and FSI^- when the two anions are present in the electrolyte as salt or in the ionic liquid. This result can partially explain why adding FSI^- in the electrolyte prevents the $EMIM^+$ co-intercalation in the graphite by forming a favorable SEI during the first reduction. Moreover, this study sheds light on the stability of the ionic liquid electrolyte under the X-ray source, which can limit our ambition in performing *operando* XPS measurements at the interface of ionic liquid electrolytes versus working electrodes, [16],[14] where alternative solutions need to be invented for such an issue. Finally, to overcome the electrolyte instability when investigating the Li^+ ion solvation, we recommend the use of liquid jet [15] where fresh electrolyte is continuously measured.

References

1. D. Liu, et al., Advanced Materials **31**, 1806620 (2019)
2. J.G. Connell, IOP Publishing book chapter, 8 (2021)
3. L. Bodenes, R. Dedryvère, H. Martinez, F. Fischer, C. Tessier, and J.-P. Pérès, Journal of The Electrochemical Society **159**, A1739 (2012)
4. K.N. Wood and G. Teeter, ACS Applied Energy Materials **1**, 4493 (2018)
5. S. Wi, V. Shutthanandan, B.M. Sivakumar, S. Thevuthasan, V. Prabhakaran, S. Roy, A. Karakoti, and V. Murugesan, Journal of Vacuum Science & Technology A **40**, 010808 (2022)
6. I. López, J. Morey, J.B. Ledeuil, L. Madec, and H. Martinez, Journal of Materials Chemistry A **9**, 25341 (2021)
7. J. Auvergniot, A. Cassel, J.-B. Ledeuil, V. Viallet, V. Seznec, and R. Dedryvère, Chemistry of Materials **29**, 3883 (2017)
8. S. Malmgren, K. Ciosek, M. Hahlin, T. Gustafsson, M. Gorgoi, H. Rensmo, and K. Edström, Electrochimica Acta **97**, 23 (2013)
9. G. Assat, D. Foix, C. Delacourt, A. Iadecola, R. Dedryvère, and J.-M. Tarascon, Nature Communications **8**, 2219 (2017)
10. X. Wu, C. Villevieille, P. Novák, and M. El Kazzi, Physical Chemistry Chemical Physics **20**, 11123 (2018)
11. X. Wu, M. Mirolo, C.A.F. Vaz, P. Novák, and M. El Kazzi, ACS Applied Materials & Interfaces **13**, 42670 (2021)
12. M. Favaro, B. Jeong, P.N. Ross, J. Yano, Z. Hussain, Z. Liu, and E.J. Crumlin, Nature Communications **7**, 12695 (2016)
13. A. Benayad, J.E. Morales-Ugarte, C.C. Santini, and R. Bouchet, The Journal of Physical Chemistry A **125**, 1069 (2021)
14. D. Weingarth, A. Foelske-Schmitz, A. Wokaun, and R. Kötz, Electrochemistry Communications **13**, 619 (2011)
15. M. El Kazzi, I. Czekaj, E. J. Berg, P. Novák, and M.A. Brown, Topics in Catalysis **59**, 628 (2016)
16. D. Streich, P. Novák, and M. El Kazzi, ECS Meeting Abstracts **MA2015-02**, 427 (2015)

Localized and structured growth of polymer brushes using ink jet printing approach

Tarek Ali Azouaou[1], *Justin* Lemarchand[1], *Vincent* Noël[1], *Giorgio* Mattana[1], *Philippe* Decorse[1] and *Christian* Perruchot[1*]

[1] Université Paris Cité, CNRS, ITODYS, F-75013 Paris, France

Abstract. Inkjet printing (IJP) technique has been used for the deposition of silica nanoparticles with controlled position and shape onto flexible polyethylene naphthalate (PEN) substrate. The printed silica areas have been chemically functionalized with a silane molecule bearing an ATRP initiator group. Surface-initiated atom transfer radical polymerisation (SI-ATRP) methodology has been betrothed to grow polymer brushes onto the silica printed areas. X-ray photoelectron spectroscopy (XPS) has been used after each step of functionalisation in order to determine the elemental surface composition. In addition, IJP has been used to design specific pattern at 250 micrometer scale. By combining maskless IJP technic and XPS line-scan mode, this work clearly demonstrates that the position and the pattern shape of printed areas can be controlled on demand, and that, structurally polymer brushes are only generated onto the functionalised silica printed areas.

Introduction

Surface modification of conducting or insulating substrates by polymer brushes has become a powerful approach to tailor on demand either the chemical or physical properties of material interfaces [1]. To ensure long-term stability and durability of the properties, each polymer brush must be covalently tethered by one chain-end to the substrate surface. Among the various approaches to graft polymers onto surface substrate, surface-initiated controlled radical polymerisation (SI-CRP) has demonstrated to be very powerful and allows great advances for surface engineering. Indeed, polymer brushes have several applications, mainly in the fields of stimuli-responsive surfaces (pH-, thermo- sensitive), biofouling surfaces, bio-compatible surfaces, bio-sensing surfaces, wetting / non-wetting surfaces [2-3]. SI-CRP approach is mostly employed as a result of its relative simple experimental setup, using mild temperature conditions, high tolerance towards a large variety of functional monomers and is also compatible with aqueous or green chemistry media. In addition, good control of the thickness of the top layer polymer can be obtained, with a high grafting density. However, all SI-CRP strategies need pre-treatment of the surface substrate with specific functional group that can initiate the polymerisation reaction to process. Among the grafting from strategies, surface-initiated atoms transfer radical polymerisation (SI-ATRP) has been widely used since halogenated initiator moieties can be easily grafted onto the surface substrate [4-5]. SI-ATRP allows the growing of polymer brushes on various organic, inorganic, metallic or semi-conducting substrates [6]. Grafting of the initiator group is obtained by chemical or electro-chemical treatment by dipping the substrate into a solution containing a bi-functional molecule (i.e. molecule containing a reactive anchoring group on one end and an ATRP-initiator group at the other end). The entire surface of the substrate is then covered by the ARTP-initiator moieties and the polymer brushes grow from the complete treated surface. Consequently, there is no control of the position or shape of the polymer area onto the substrate surface.

In the recent years, inkjet printing (IJP) technique has attracted considerable attention since easy functionalisation of planar and flexible substrate can be obtained using commercially available inks formulation containing either inorganic (i.e. silica) or metallic (i.e. copper, silver, gold) nanoparticles suspension in aqueous media [7]. IJP is very a versatile and adjustable technique since the position, the shape and the dimension (from ~100µm up to centimeter length) of the printed area can be easily controlled via dedicated PC software. IJP is a fully digital and non-contact technique, but more importantly, it is a maskless technique. In addition, IJP is performed at ambient conditions and do not require large volumes of ink. Using IJP technique, the dimension of the printed area fill the gap between the fully treated-surface using conventional chemical, electrochemical or physical (plasma, CVD, PVD) approaches and functionalisation of nanometer-size area scales using lithography (photo- or electro-) or micro-electrochemistry techniques [8].

In this work, inkjet printing technique has been used for the deposition of nanoparticles of silica (i.e. *NP-silica*) with controlled position and shape onto flexible

* Corresponding author: christian.perruchot@u-paris.fr

polyethylene naphthalate (PEN) substrate. A silane molecule, functionalised with a halogenated ATRP end-group, is then covalently grafted onto the silica printed areas. Finally, SI-ATRP procedure, using different methacrylate monomers, has been used to grow polymer brushes onto the silica printed areas. X-ray photoelectron spectroscopy (XPS) has been used to control the modifications of the elemental surface composition after each chemical treatment of the substrate, and to confirm the growth of polymer brushes onto the printed areas. In addition, onto patterned surface, XPS line-scan mode allows to reconstruct the surface composition.

Experimental

Ink jet printing:

Poly(ethylene) naphthalate (PEN Teonex® Q65HA) flexible and transparent substrate was purchased from Putz Folien (Germany). Fumed NP-silica (Sigma) was used to formulate the ink suspension (1% wt / vol in ethanol-water 90-10 vol / vol). Areas functionalised by silica were inkjet-printed using a Fujifilm Dimatix Materials Printer DMP-2850 [9], directly on the PEN substrate. The ejected volume of the dropped ink was 10 pL and the drop spacing was set at 10 µm. The printer platen temperature was maintained at 40°C during printing, to promote rapid evaporation of solvent and reduce the 'coffee-ring' edge effect. Finally, the printed pattern areas were thermally sintered in an oven at 180°C during one hour.

ATRP initiator:

3-(2-bromoisobutyramido) propyltriethoxysilane (chemical structure in Figure 1a) has been synthesized according to literature [10]. The silane functionalized with the ARTP initiator was grafted by immersing the PEN-silica substrate in a water/ethanol solution (80 / 20 vol. by vol.) containing 1% silane molecule for 6 hours. The resulting functionalized substrate was carefully and abundantly rinsed with ethanol to remove all excess of unreacted silane, then air-dried overnight at room temperature (RT). The silane end group acts as anchoring species to silica printed area whereas the halogenated end groups serve as initiator for the ARTP process in order to grow the polymer brushes.

ATRP Chemicals and procedure:

Copper bromide (CuBr) catalyst, pentamethyldiethyl triamine (PMDETA) ligand and methacrylate monomers (hydroxyethyl methacrylate (HEMA) and pentafluoropropyl methacrylate (PFPMA)) were purchased from Sigma Aldrich and used without further purification. The chemical structure of HEMA and PFPMA monomers are presented in Figures 1b and 1c respectively.

In a typical polymerisation procedure, the functionalised initiator grafted onto silica printed area was immersed into argon-degassed acetonitrile solution (6 ml) containing CuBr catalyst (100 mg, 0.68 mmol) PMDETA ligand (150 µL, 0.68 mmol) and methacrylate monomer (1.66 mL for HEMA or 2.32 ml for PFPMA). The catalyst / ligand / monomer were set at 1 / 1/ 20 molar ratio for all experiments. The reaction polymerisation was moderately stirred overnight at RT. The resulting material was carefully rinsed with acetonitrile to remove all non-grafted catalyst/ligand and monomer species. The polymer brushes grafted onto the silica printed area were vacuum-dried prior to XPS analysis. The sample notation is as follows:

PEN-SiO_2-Init-pHEMA (for HEMA monomer) or PEN-SiO_2-Init-pPFPMA (for PFPMA monomer).

a

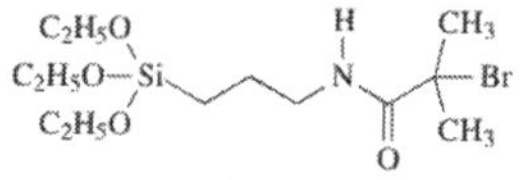

b

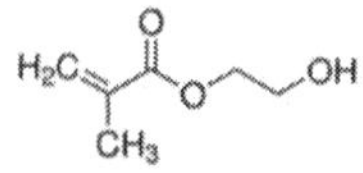

c

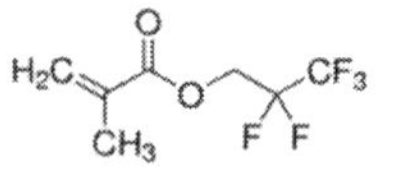

Fig. 1: Chemical structure of (a) silane bearing ATRP initiator, (b) HEMA and (c) PFPMA molecules

XPS Characterization:

XPS analysis were performed on a K Alpha$^+$ spectrometer (ThermoFisher), equipped with a monochromated X-Ray Source (AlKα, 1486.6 eV). An X-ray spot size of 400 µm was used. The hemispherical analyzer was operated in CAE (Constant Analyzer Energy) mode, with a pass-energy of 100 eV and a step of 1 eV for the acquisition of survey spectra, while a pass-energy of 40 eV and a step of 0.1 eV were used for the acquisition of high resolution spectra. The spectra were recorded and data processing was performed with Advantage© software (V 5.99 from ThermoFisher). The elemental surface compositions were calculated by determining the area of each element of interest with a Shirley type background subtraction and using the Scofield sensitivity factors of the spectrometer. The spectra were calibrated against the C_{1s} main peak component C-C/C-H set at 285.0 eV.

Results and discussion

Inkjet printed areas are digitally controlled via appropriate software, but more importantly, IJP is a maskless technique. Thus the position and the pattern, with shape and dimension varying from *ca* 100 µm lateral resolutions up to centimeter scale, of the printed area can be controlled on demand.

In a preliminary experiment, IJP was used for the deposition of silica on 0.5 * 0.5 mm^2 areas onto the PEN substrate.

Figure 2 summarizes the stepwise strategy used to grow polymer brushes onto the PEN substrate. The three steps are (i) surface modification of PEN substrate by inkjet printing of silica patterns, (ii) grafting the silane functionalised with the ATRP initiator and (iii) the growth of the polymer brushes using ATRP procedure.

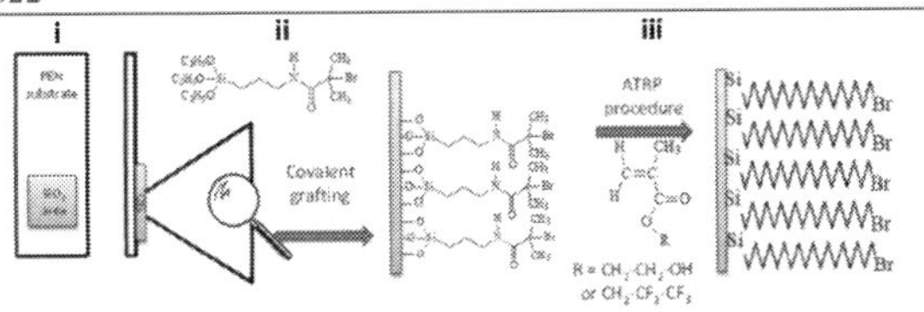

Fig. 2: General scheme summarizing the stepwise strategy used for surface modification of PEN substrate by (i) inkjet printing of silica pattern (0.5 * 0.5 mm^2), (ii) grafting the silane functionalised with the ATRP initiator and (iii) the growth of the polymer brushes using ATRP procedure.

XPS was used to monitor the changes in the surface chemical composition of the PEN substrate after each functionalization step [11]. Figure 3 displays the survey spectra of bare PEN substrate (Fig. 3a), silica inkjet-printed area (Fig. 3b), silica chemically functionalized with the silane bearing the ARTP initiator (Fig. 3c), and after the growth of polymer brushes using HEMA (Fig. 3d) or PFPMA (Fig. 3e) monomer by ATRP procedure.

For bare PEN substrate (Fig. 3a), carbon C_{1s} (285 eV) and oxygen O_{1s} (532 eV) peaks are clearly detected, as expected. Adventitious nitrogen N_{1s} (400 eV) and calcium Ca_{2p} (349 eV) peaks are weakly detected.

After IJP of a silica layer (Fig. 3b), the relative intensity of carbon C_{1s} peak is decreasing whereas the relative intensity of oxygen O_{1s} peak is increasing (oxygen arising from PEN substrate and silica top layer). Additional peaks of silicon Si_{2p} (102 eV) and Si_{2s} (153 eV) are clearly detected. These latter peaks are characteristic of the silica deposited surface layer.

Chemical grafting of the silane bearing ATRP initiator (Fig. 3c) is clearly evidenced by the detection of nitrogen N_{1s} (~400 eV) and bromine Br_{3d} (~70 eV) - Br_{3p} (182-190 eV) doublet peaks respectively. These elements are unambiguous markers and clearly demonstrate the effective grafting of the ATRP initiator groups onto the silica layer. Silane derivatives are molecules known to react easily towards hydroxyl surface groups present on metallic, oxide or inorganic (alumina, silica) surfaces [12].

Using the ATRP procedure with methacrylate monomer, polymer brushes are generated onto the silica areas. In the case of HEMA monomer (Fig 3d), carbon C_{1s} and oxygen O_{1s} peaks are clearly observed, whereas the relative intensities of silicon Si_{2p} and Si_{2s} peaks considerably decrease. Nitrogen N_{1s} signal is almost not detected. These observations clearly demonstrate that the outermost pHEMA brushes mask the underlying silica layer, inducing a significant decrease of the silicon and nitrogen peaks intensity [13].

When PFPMA monomer is used to grow polymer brushes, in addition to carbon C_{1s} and oxygen O_{1s} peaks, silicon Si_{2p} and Si_{2s}, nitrogen N_{1s} and bromine Br_{3d} signals are still detected, but with lower intensities when compared to Figure 3c. However additional peaks assigned to fluorine F_{1s} (~689 eV) and F_{Auger} (~835 eV) are clearly detected. Fluorine is an unambiguous elemental marker of PFPMA monomer (5 fluorine atoms per monomer unit). Thus the detection of fluorine peaks are a direct confirmation that polymer brushes are grafted onto the silica layer.

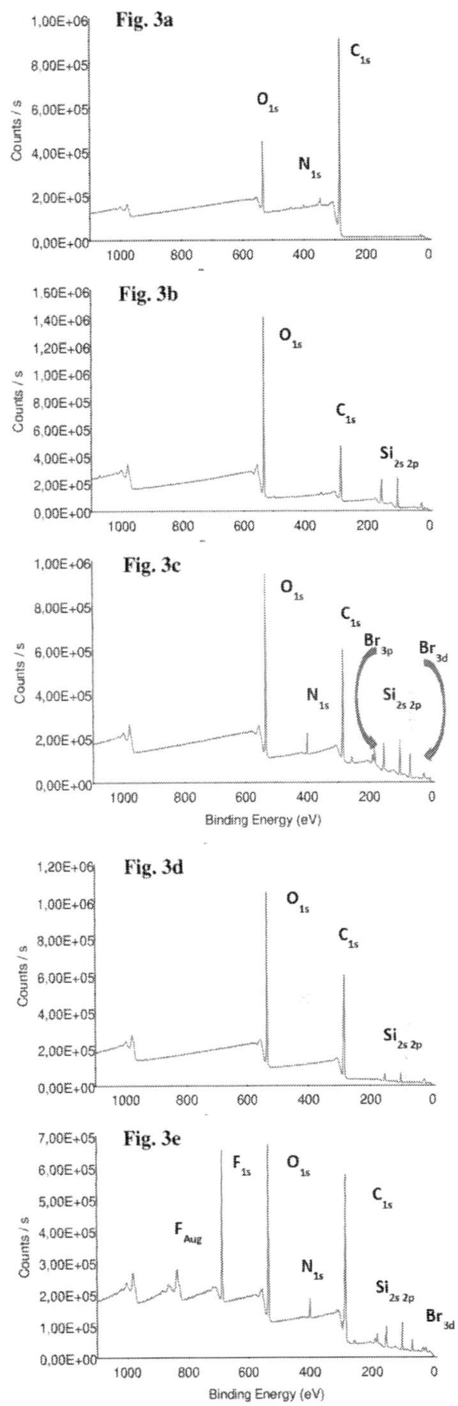

Fig. 3: XPS survey spectra of (a) bare PEN substrate, (b) silica IJP area, (c) silica functionalized with the silane bearing the ARTP initiator, polymer brushes using (d) HEMA or (e) PFPMA monomer.

The surface elemental compositions as determined by XPS of the PEN substrate after each step of functionalization are reported in Table 1.

Element (%) / Sample	C	N	O	Si	Br	F

Raw PEN	83.5	1.1	15.0	0.4	---	---
PEN-SiO$_2$	36.7	0.1	44.5	18.7	---	---
PEN-SiO$_2$-Init	47.1	4.8	31.4	14.4	2.4	---
PEN-SiO$_2$-Init-p(HEMA)	63.1	0.4	31.7	4.8	< 0.1	---
PEN-SiO$_2$-Init-p(PFPMA)	55.1	3.3	22.4	7.3	0.9	11.0

Table 1: Surface elemental compositions as determined by XPS.

When compared to bare PEN substrate, the deposition of a thin silica layer by IJP onto the PEN substrate induced a significant decrease of the carbon content, and an increase of the oxygen and silicon contents as expected. The PEN substrate is partially masked by the top silica layer. Grafting of the silane bearing the ATRP initiator group is evidenced by the increase of the nitrogen and bromine contents. XPS analyses clearly demonstrate that polymer brushes are present at the outermost surface of the silica layer. Indeed, the significant decrease of nitrogen and silicon contents (for HEMA monomer) and the specific detection of fluorine peaks (for PFPMA monomer) are unambiguous proof that the ATRP initiator is still active after chemical grafting and that it can engender the growth of polymer brushes via SI-ATRP procedure. The outermost polymer brushes thus mask the underlying PEN-silica substrate [14].

In order to push even further the lateral resolution of the IJP technique, four strips of silica, having a width of 250 µm and apart by 300 µm, have been patterned onto the PEN substrate. Figure 4 presents an optical view (partial view recorded from the Dimatix printer camera) of the resulting silica bands deposited by IJP.

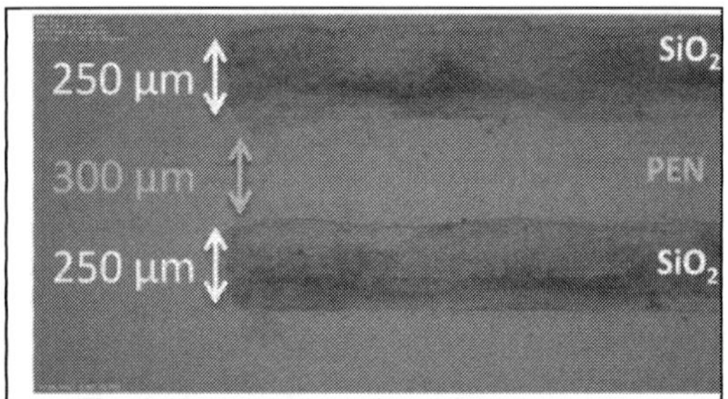

Fig. 4: Partial optical view of the resulting silica strips (250 µm width and apart by 300 µm).

As previously described, similar protocols have been employed in order to (i) graft the silane functionalized ARTP-initiator and (ii) to growth polymer brushes with PFPMA monomer using ATRP procedure, onto this patterned PEN-Silica surface.

The resulting patterned material has been characterized by XPS using the line-scan mode. The diameter of the X-ray source was set at 100 µm and the step size was fixed at 50 µm. The line-scan analysis was performed perpendicularly to the silica strips pattern. Figure 5 presents the reconstructed carbon, silicon and fluorine contents as determined by scan-line mode analysis.

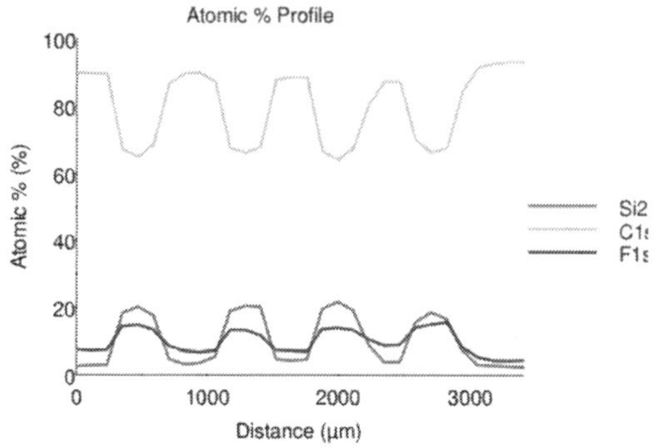

Fig. 5: Plot of the reconstructed carbon, silicon and fluorine contents as determined by scan-line mode analysis.

The scan line mode analysis clearly shows a "square wave" form. The carbon rich areas mainly arise from the uncoated PEN substrate, whereas the silicon rich areas are a direct evidence of the silica bands deposited by IJP. Silicon exhibits four maximum plateau values corresponding to the four printed silica regions. It is important to note that carbon and silicon rich areas are opposed, demonstrating that the top silica printed layer partially mask the underlying PEN substrate. The width of the maximum silicon plateau contents is approximately 250 µm, in good agreement with the width of the silica printed areas. In addition, fluorine is used as a unique elemental marker of the PFPMA polymer brushes. It is worth noticing that the shape of the fluorine content matches that of the silicon one. This observation clearly demonstrates that polymer brushes are mainly present onto the silica regions, as expected. This scan line analysis demonstrates that silica bands can be printed onto the PEN substrate and that after chemical treatment via the silane containing the ATRP initiator group, the growth of polymers brushes can be localised on the desired position with specific shape.

Conclusion

Inkjet printing technology has been used to print silica areas onto flexible PEN substrate. The silica was subsequently chemically modified with a silane molecule bearing an ATRP initiator. Then, the initiator groups are active to generated polymer brushes via SI-ATRP process on the treated areas. These results demonstrate that structurally and controlled design, of various shape and dimension, can be easily generated by inkjet printing technique. The physico-chemical surface properties of the resulting material are thus governed by the specific properties of the polymer brushes. Work is currently in progress in order to lower the dimension of the printed silica zones, to extend this approach to others flexible insulating (i.e. Kapton ®, polyethylene imine) or conductive (ITO coated PET) substrate and to pattern the substrate with more complex shapes. In addition, using others (meth)acrylate monomers bearing specific lateral group, polymer brushes exhibiting specific chemical properties will be generated only on the desired position. Thus the resulting material will exhibit specific properties only on the structurally and localised printed areas.

Acknowledgements
ANR (Agence Nationale de la Recherche) and CGI (Commissariat à l'Investissement d'Avenir) are gratefully acknowledged for their financial support of this work through Labex SEAM (Science and Engineering for Advanced Materials and devices) ANR-11-LABX-086 and ANR-11-IDEX-05-02. XPS Spectrometer was also supported by the Ile-de-France regional council, SESAME agreement n °16016303.

References

1 J.-O. Zoppe, N. C. Ataman, P. Mocny, J. Wang, J. Moraes, H.-A. Klok, Chem. Rev., 117 (2017) 1105-1318.
2 O. Azzaroni, J. Polym. Sci.: Part A: Polym. Chem., 50 (2012) 3225-3258.
3 M. Krishnamoorthy, S. Hakobyan, M. Ramstedt, J. E. Gautrot, Chem. Rev., 114 (2014) 10976-11026.
4 K. Matyjaszewski, J. Xia, Chem. Rev., 101 (2001) 2921-2990.
5 W. A. Braunecker, K. Matyjaszewski, Prog. Polym. Sci., 32 (2007) 93-146.
6 F.J. Xu, K.G. Neoh, E.T. Kang, Prog. Polym. Sci., 34 (2009) 719-761.
7 S. H Ko, H. Pan, C. P Grigoropoulos, C. K Luscombe, J. M. J. Fréchet, D. Poulikakos, Nanotechnology, 18 (2007) 345202.
8 J. Lemarchand, N. Bridonneau, N. Battaglini, F. Carn, G. Mattana, B. Piro, S. Zrig, V. Noël, Angew. Chem. Int. Ed., 61 (2022) e202200166.
9 Dimatix Materials Printer DMP-2850 - Fujifilm [United States], https://www.fujifilm.com/us/en/business/inkjet-solutions/ inkjet-technology-integration/dmp-2850.
10 T. von Werne, T.E. Patten, J. Am. Chem. Soc., 123 (2001) 7497-7505.
11 G. Beamson, D. Briggs (Editors) in « High resolution XPS of organic polymers. The Scienta 300 database ». Chichester, John Wiley & Sons, 1992.
12 "Silanes and Other Coupling Agents", Volume 3, Edited by Kash L. Mittal, Published by CRC Press, 2004.
13 C. Perruchot, A.M. Khan, A. Kamitsi, S. P. Armes, T. von Werne, T.E. Patten, Langmuir, 15 (2001) 4479-4481.
14 C. Perruchot, M.A. Khan, A. Kamitsi, S.P. Armes, J.F. Watts, T. von Werne, T.E. Patten, Eur. Polym. J., 40 (2004) 2129–2141.

Gas phase Photoemission studies in the hard X-ray domain

Marc Simon[1*]

[1]Sorbonne Université, CNRS, Laboratoire de Chimie Physique-Matière et Rayonnement, LCPMR, 75005 Paris, France

Abstract. Recent results obtained in gas phase photoemission studies are presented in this article with particular emphasis on recoil, Double Core Hole and Post Collision Interaction. These three important effects are not specific to the gas phase and could have more general applications in condensed matter studies.

1 Introduction

Light-Matter interaction studied by photoemission on gas phase samples has proven to furnish a wealth of information. This article deals with Hard X-ray Photoelectron Spectroscopy (HAXPES) with a focus on the tender x-ray domain (2-12 keV). One of the interests of this photon energy range is the short (femtosecond or less) core-hole lifetime of the reached electronic states by photoabsorption. Among the different studies performed by our group, I will select three important fundamental processes, which could attract the condensed phase community participating in the JNSPE: high recoil effects due to the high kinetic energy of the emitted photoelectron, Double Core Hole Spectroscopy and Post Collision Interaction. These studies have important applications like radiation-damage in the hard x-ray domain. The measurements were performed at SOLEIL, the French Synchrotron facility, located close to Paris. The perspectives offered by these results are discussed. We have published several review articles in the past where the reader could find more details [1-3].

2 Experimental setup

The HAXPES experimental chamber is a permanent endstation of the SOLEIL GALAXIES beamline.

2.1 The SOLEIL GALAXIES beamline

The GALAXIES beamline is an undulator beamline, optimized to operate in the 2-12 keV energy range [4]. The undulator beam is monochromatized by a double-crystal monochromator providing a photon flux of few 10^{12} photons per second on the sample. A 4-bounce, high-resolution monochromator (HRM) has been recently commissioned allowing to obtain a photon bandwidth of 100 meV between 4 and 12 keV.

2.2 The HAXPES endstation

HAXPES is a permanent endstation of the GALAXIES beamline [5]. A high-resolution hemispherical electron analyzer allows to measure electron spectra with a kinetic energy resolution of 35 meV for electron kinetic energies up to 12 keV. For gas phase studies, a gas cell is installed into the photon beam thanks to a motorized 4-axes manipulator.

*Corresponding author: marc.simon@sorbonne-universite.fr. ORCID: https://orcid.org/0000-0002-2525-5435

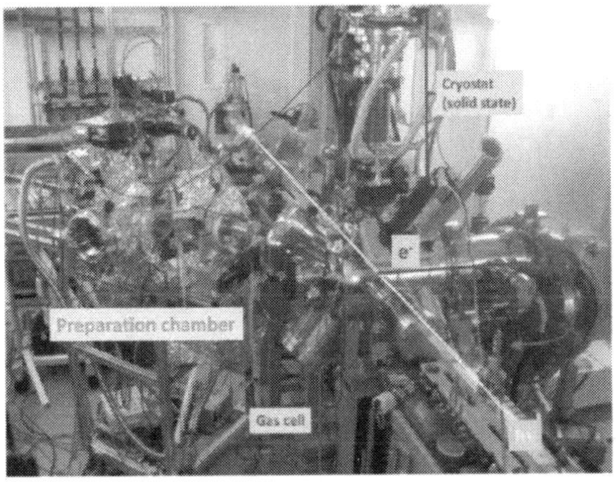

Fig. 1. Photo of the HAXPES endstation of the SOLEIL GALAXIES beamline.

3 Recoil effects

In the hard x-ray domain, the momentum of the departing photoelectron is far to be negligeable as it is the case in the soft x-ray domain. This leads to translational recoil for isolated atoms and molecules. Additionally, depending on the direction in which the photoelectron is ejected, vibrational and rotational recoil have been observed for molecules.

3.1 Translational recoil

Core ionization of neon has been studied as a demonstration case [6]. Ne1s ionization has been performed at high photon energy and the Auger electrons have been measured for different photon energies. The recoil momentum of the Ne^+ ion is so high that Atomic Auger Doppler effects have been observed, as illustrated in figure 2.

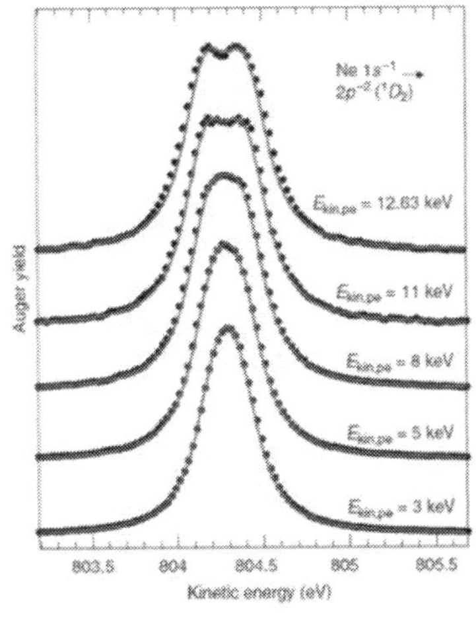

Fig. 2. Atomic Auger Doppler effects of core-ionized neon at various kinetic energy of the 1s photoelectron.

3.2 Vibrational recoil

In the case of carbon 1s photoionization of the CF_4 molecule, the escaping C1s photoelectron cannot induce rotational recoil of the CF_4^+ ion because the carbon atom is located at the center of gravity of the tetrahedral molecule. In addition to the translational recoil, only vibrational recoil is then induced by the fast C1s photoelectron emission, as illustrated in figure 3 [7]. Close to the C1s Photoionization threshold, only a single vibrational level of the CF_4^+ ion is populated. When the photon energy is increased, a wider vibrational envelope is observed as a signature of the vibrational recoil.

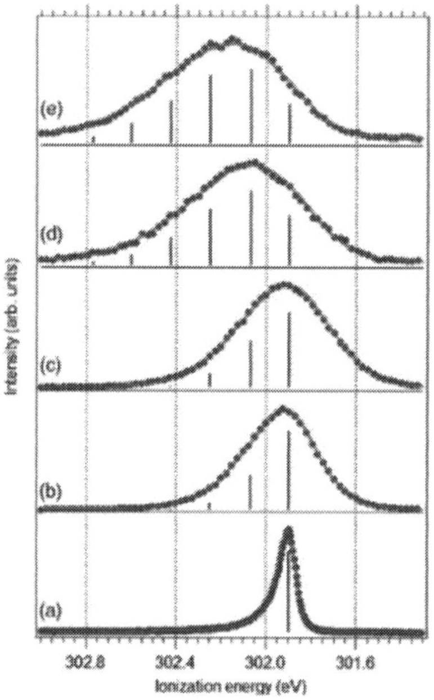

Fig. 3. Vibrational recoil of the C1s core-ionized CF_4 molecule measured at different photon energies.

3.3 Rotational recoil

In case of the CO molecule, C1s photoionization can lead to vibrational recoil if the photoelectron is emitted parallel to the molecular axis and to rotational recoil if the photoelectron is emitted perpendicular to the molecular axis. Rotational levels cannot be resolved at such high kinetic energies. Rotational recoil effects are observed as a shift of the centroid of the photoelectron peak toward lower kinetic energy, the difference in energy being the rotational energy of the molecular ion. This rotational recoil has been precisely determined by a mixture of CF_4 and CO molecules as illustrated in figure 4.

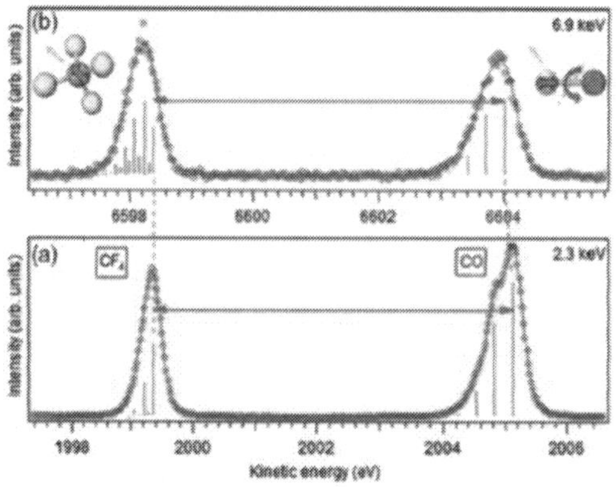

Fig. 4. C1s photoelectron of a mixture of CF_4 and CO molecule. Arrows are showing the kinetic energy difference between the v=0 of both molecular ions which vary with the upcoming photon energy due to the rotational recoil.

3.4 Recoil induced ultrafast molecular rotation of the CO molecule probed by dynamical rotational Doppler effect

Photoionization at high photon energy induces ultra-fast rotation up to an effective rotational temperature of about 10,000 K due to the recoil caused by emission of a fast photoelectron. In order to monitor the recoil-induced rotational motion of the CO molecule, high-resolution Auger spectra of the C1s core-ionized molecular ion have been recorded at various photon energies [8]. Given the C1s core-hole lifetime, the molecular ion has a time of 5 femtoseconds to rotate before the Auger electron emission. This is illustrated in figure 5.

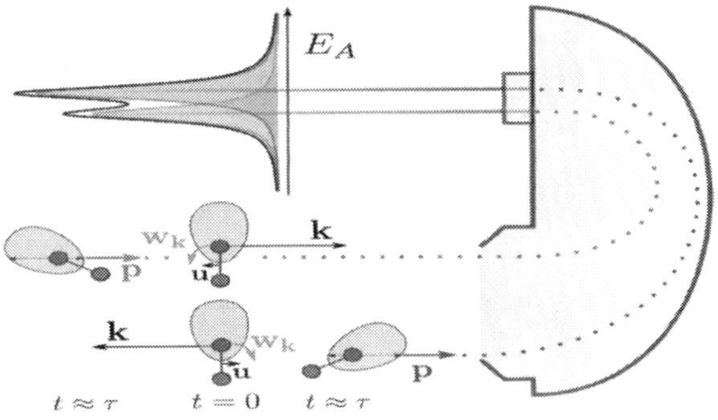

Fig. 5. Classical picture of the dynamical rotational Doppler effect of the C1s core-ionized CO molecule

4 Double Core Hole Spectroscopy

Double Core Hole (DCH) studies are attract considerable interest mainly because of the high chemical sensitivity of the DCH binding energy compared to the traditional ESCA (Electron Spectroscopy for Chemical Analysis) technique [9]. In case of ESCA, a chemical shift of 700 meV has been recently measured between the CH_3I and CF_3I molecules at the iodine 1s ionization threshold around 33 keV binding energy [10]. These results have been obtained at the Spring8 synchrotron radiation facility where an HAXPES endstation has been successfully commissioned in the hard x-ray domain [11]. The high sensitivity of DCH states has been theoretically predicted during the 1980's and but only experimentally confirmed very recently [12]. Most of the DCH studies have used sophisticated magnetic bottles in order to measure both core-ionized photoelectrons in a coincidence regime. An interesting way to study DCH states has been recently founded [13]: one of the core electrons is ejected into the continuum whereas a second core electron is promoted to an empty molecular orbital.

4.1 Spectroscopy of DCH states

The high-resolution of the HAXPES setup is particularly well adapted to perform DCH studies with one core electron ionized and the other promoted into an unoccupied level. A proof of principle has been performed with the argon $1s^{-1}2p^{-1}$ shake-up photoelectron spectrum shown figure 6 [14]. The Rydberg series of the DCH states converging towards the $^3P_2$, $^3P_1$, $^3P_0$ and $^1P_1$ DCH thresholds are clearly visible with a good signal/noise ratio. Spectral bandwidth has been measured allowing to determine the lifetime of the DCH states.

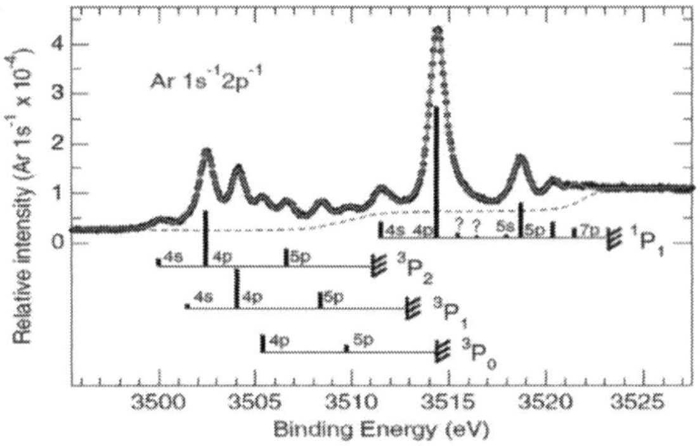

Fig. 6. $1s^{-1}2p^{-1}$ DCH shake-up photoelectron spectrum of argon

This technique has been applied to the neon atom [15], the H_2O molecule [16], the CS_2 and SF_6 molecules [17], the HCl molecule [18], the O K^{-2} CO_2 molecule [19].

4.2 Electronic relaxation of DCH states

The Auger relaxation of these DCH states mainly occurs via a 2 step Auger decay: a first Auger decay filling one core hole of the DCH state, called hypersatellite Auger decay, occurs followed by a second Auger decay filling the remaining core hole. Even if the relative ratio of DCH/SCH states is in the order of 10^{-3}, good signal/noise ratio spectra have been obtained on neon [15, 20] as illustrated in figure 7.

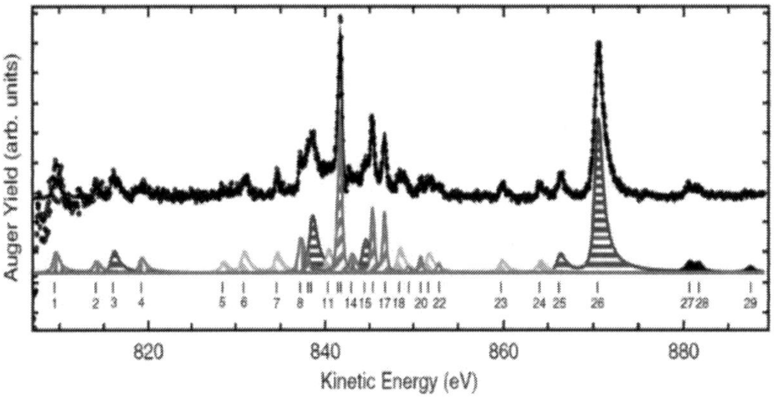

Fig. 7. Hypersatellite Auger spectrum of neon.

The relaxation of these DCH states mainly occurs via a 2 step Auger decay: a first Auger decay filling one core hole of the DCH state, called hypersatellite Auger decay, occurs followed by a second Auger decay filling the remaining core hole. Even if the relative cross section of DCH/SCH states is in the order of 10^{-3}, nice signal/noise ratio spectra have been obtained.

The hypersatellite Auger spectrum of the H_2O molecule has revealed [21] ultra-fast nuclear motion occurring within the lifetime of the DCH state which is equal to 1.5 femtosecond.

4.3 DCH states studies through cascade Auger decay

Deep shell excitation leads to cascade Auger decay. Resonant KLL Auger decay after resonant K shell excitation of the HCl molecule have revealed ultra-fast fragmentation of the HCl^+ ion prior to the Auger decay of the L^{-2}DCH state [22-23] as illustrated in figure 8. The gradient of the Potential Energy Curve is so strong that the wave packet propagation has time to propagate during the 3 femtoseconds lifetime of the L^{-2} DCH state.

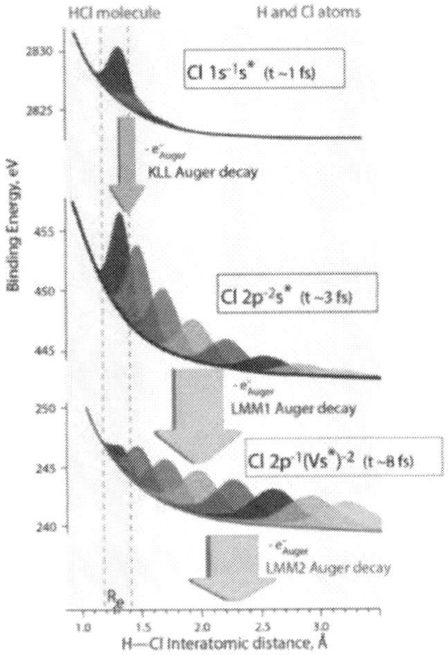

Fig. 8. Potential energy curves for different steps of the KLL Auger cascade following Cl $1s \rightarrow \sigma^*$ excitation of the HCl molecule.

Complementary Resonant Auger and Resonant X-ray emission measurements allowed to determine the lifetime of the DCH state and the gradient of the Potential Energy curve of the CH_3I molecule [24].

5 Post Collision Interaction (PCI)

Post Collision Interaction (PCI) effect takes place in presence of a slow photoelectron and a fast Auger electron. Close to the ionization threshold, the slow photoelectron starts to leave a singly charged ion, but then the faster Auger electron overtakes it. Then, instead of "feeling" the potential of a singly charged ion, the photoelectron suddenly 'feels' a doubly charged ion. Due to PCI, the photoelectron loses kinetic energy and the Auger electron simultaneously gains kinetic energy [25-27]. Just above threshold, the photoelectron can be recaptured into a unoccupied Rydberg orbital. Momentum transfer between the slow photoelectron and the fast Auger electron can also takes place [28].

PCI is a quite general effect. Several groups are working for the moment on PCI on liquids and solid states compounds.

6 Conclusion

All the presented effects presented in this overview have been observed in the gas phase during the last ten years. Recoil, Double Core Hole and PCI are easier to observe with diluted species but can be studied in any phases.

Acknowledgements

I warmly acknowledge all the coauthors of the presented articles and particularly Maria Novella Piancastelli, Oksana Travnikova, Tatiana Marchenko, Renaud Guillemin, Iyas Ismail, Ralph Püttner and Denis Céolin.

References

1. M. Simon, M. N. Piancastelli and D. W. Lindle *Hard X-Ray Photoelectron Spectroscopy from Atoms and Molecules (HAXPES)* (*Springer Series in Surface Science* vol 59) ed J. Woicik (New York: Springer) (2016)
2. M. N. Piancastelli et al., J. Phys. B 50 042001 (2017)
3. M. N. Piancastelli et al., Rep. Prog. Phys. 83 016401 (2020)
4. J.-P. Rueff et al., J. Synchrotron Radiation 22 175 (2015)
5. D. Céolin et al., J. Electron Spectrosc. and Relat. Phenom. 190 188 (2013)
6. M. Simon et al., Nature Communications 5 4069 (2014)
7. E. Kukk et al., Phys. Rev. Lett. 121 073002 (2018)
8. D. Céolin et al., Proceedings of the National Academy of Sciences 116 4877 (2019)
9. L. Cederbaum et al., J. Chem. Phys. 85 6513 (1986)
10. N. Boudjemia et al., Phys. Chem. Chem. Phys. 21 5448 (2019)
11. M. Oura et al., New J. Phys. 21 043015 (2019)
12. D. Koulentianos et al., J. Chem. Phys. 149, 134313 (2018)
13. M. Nakano et al., Phys. Rev. Lett. 111 123001 (2013)
14. R. Püttner et al., Phys. Rev. Letters, 114, 093001 (2015)
15. G. Goldsztejn et al., Phys. Rev. Letters 117, 133001 (2016)
16. T. Marchenko et al., J. Physics B 53 224002 (2021)
17. R. Feifel et al., Scientific Reports 7 13317 (2017)
18. D. Koulentianos et al., Phys. Chem. Chem. Phys. 20, 2724 (2018)
19. D. Koulentianos et al., Phys. Chem. Chem. Phys. 23, 10780 (2021)
20. G. Goldsztejn et al., Phys. Rev. A 96 012513 (2017)
21. T. Marchenko et al., Phys. Rev. A 98 063403 (2018)
22. O. Travnikova et al., Physical Review Letters116, 213001 (2016)
23. O. Travnikova et al., Physical Review Letters 118, 213001 (2017)
24. T. Marchenko et al., Physical Review Letters 119, 133001 (2017)
25. R. Guillemin et al., Physical Review Letters 109, 013001 (2012)
26. R. Guillemin et al., Physical Review A 92, 012503 (2015)
27. S. Kosugi et al., Physical Review Letters 124, 183001 (2020)
28. R. Guillemin et al., Physical Review A 99, 063409 (2019)

ToF-SIMS and XPS characterizations of model fission products (I, Cs) deposits after thermal treatment simulating late phase conditions of a nuclear power plant severe accident.

LE FESSANT Elouan[1], GREGOIRE Anne-Cécile[1], NUNS Nicolas[3], SIMON Pardis[2], MAMEDE Anne-Sophie[2], PAUL Jean-François[2], CANTREL Laurent[1]
[1] *Institut de Radioprotection et de Sûreté Nucléaire, PSN-RES, BP 3, 13115 Saint-Paul-lez-Durance.*
[2] *Univ. Lille, CNRS, Centrale Lille, Univ. Artois, UMR 8181 – UCCS – Unité de Catalyse et Chimie du Solide, F-59000 Lille, France*
[3] *Univ. Lille, CNRS, INRAE, Centrale Lille, Univ. Artois, FR 2638 - IMEC - Institut Michel-Eugène Chevreul, F-59000 Lille, France*

Abstract. During a Severe Accident (SA) occurring in a nuclear power plant, many Fission Products (FP) are released from the degraded fuel and are transported in the Reactor Coolant System (RCS). Depending on their volatility, FP can be either deposited on the surface of the Reactor Coolant System (RCS) or transported into the containment building from where they may be released into the environment in case of early containment failure. This was the case for the Fukushima Daiichi (FD) accident with important FP releases which were observed following the containment vessel failure which lasted up to several weeks after the accident. If early phase releases were well predicted by SA simulation codes, delayed releases were not correctly predicted [1]. Such discrepancies can be attributed to the revaporization of deposits inside the RCS which were not yet modeled in the SA simulation tools. Previous experiments over the past 20 years evidenced the possible formation and revaporization of CsI [2]. However, up to now, no clear CsI revaporization mechanism has been determined. In order to propose a mechanism, this study (scheduled within the OECD/NEA/ESTER project) aims at determining the speciation of each species remaining on the substrate after revaporization tests under different conditions using XPS and ToF-SIMS (X-ray Photoelectron Spectroscopy and Time of Flight Secondary Ion Mass Spectrometry. These two combined techniques allow the identification of the main species of the residue leading to the proposition of two revaporization reactions.

Keywords: XPS, ToF-SIMS, Revaporization, CsI

1. Introduction

During a Severe Accident (SA) occurring in a nuclear power plant, many Fission Products (FP) released from the reactor core region under gaseous or aerosols forms can be either deposited on the surface of the Reactor Coolant System (RCS) or transported into the containment building. During the Fukushima Daiichi (FD) accident, important FP releases involving iodine and cesium were observed following the containment vessel failure and lasted for several weeks after the start of the accident. While early releases are correctly predicted by SA simulation tools, simulations of delayed releases should be improved [1]. The delayed releases observed in the case of FD can be attributed to the revaporization of deposits inside the RCS. To improve simulations, especially in the case of iodine and cesium, the revaporization behavior of CsI – which is one of the main expected deposited forms in the RCS - has been studied for several decades.

Previous experiments over the past 20 years evidenced the conditions that favor the revaporization of CsI deposits [2-3]. CsI powder revaporization is thus reported to start below 500°C and to be completed around 750°C in a steam atmosphere – the main revaporized species is CsI as vapor which can be transported by the steam flow in colder zones where it condenses as submicrometric aerosols. More recent work focused on the revaporization of CsI aerosol deposits in a temperature ramp (250-750°C)[4]. Under pure air, iodine (I) was released mainly (up to 90% of the initial amount of I) as molecular iodine (I_2) whereas, under pure steam, only CsI was vaporized. For a mixed air/steam atmosphere, revaporization results partly in the formation of gaseous I_2 and CsI, regardless of the air/steam ratio.

Up to now, no clear mechanism for CsI revaporization has been determined – especially in oxidizing conditions. In order to address this issue, our study (in the frame of the OECD/NEA/ESTER project) aims to determine the speciation of each species remaining on oxidized 304L stainless steel samples (called the substrate). After CsI revaporization under different conditions surface characterization techniques are used such as XPS (X-ray Photoelectron Spectroscopy) and ToF-SIMS (Time of Flight - Secondary Ion Mass Spectrometry). Indeed, XPS provides the chemical environment of each element present in the 10 nm surface layer and ToF-SIMS allows the identification of elemental and molecular fragments emitted from the outermost surface. These two complementary techniques allow the identification of the main species remaining at the substrate surface.

Experimentations performed in this study focused on isothermal treatments (between 200 °C and 600 °C) under an oxygen partial pressure over 10^{-2} atm. The first results highlight an increase in gaseous iodine release with temperature. The main objective of this paper is to investigate the nature of the residue, allowing then a better understanding of the mechanisms resulting in the release of gaseous species.

2. Experimental protocol

To be as representative as possible of a SA situation in the RCS, our experimental protocol consists in 4 successive steps:

1) Substrate preparation: during a SA, RCS Stainless Steel (SS) is oxidized by the temperature rise combined with steam. Hence, coupons (1.2 cm^2) of SS representative of RCS (304L) are oxidized to simulate the RCS oxidation. These pieces of oxidized SS will be used as a substrate for aerosol deposits;
2) CsI deposition: a previous study [4] showed that deposits obtained by aerosol impaction at room temperature behave as deposits obtained at higher temperatures. Thus, we retained the technique of aerosol deposition at room temperature. Aerosols are generated by nebulization-drying a CsI concentrated aqueous solution and next deposited by impaction on the SS oxidized sample;
3) CsI revaporization: deposits are vaporized in the ATMIRE experimental revaporization setup to simulate FP revaporization (an open flow reactor placed in a high-temperature furnace allowing the atmosphere control above the studied sample);
4) Characterization: Both elementary analyses (ICP-MS) and surface analyses (XPS and ToF-SIMS) are performed to determine the mass of residue and to identify its nature with possible interaction with the substrate. XPS analyses are done on a Kratos Analytical AXIS Ultra^{DLD} with a monochromatic Al Kα source at 1486.6 eV. The adventitious C 1s (285.0 eV) binding energy (BE) was used as an internal reference. ToF-SIMS analyses are performed on a ToF-SIMS V instrument (ION-TOF GmbH) with a Bi^+ beam (25 keV, 1,15 nA).

3. Experimental results

Different isotherms were investigated to highlight the temperature influence on CsI reactivity. Two levels of temperature were retained: 410 °C and 520 °C. Tests were performed under 0.8 atm steam partial pressure and 0.2 atm in air (4.10^{-2} oxygen partial pressure).

3.1 Cs speciation after revaporization at 410 °C

Remaining elements on the SS coupon were analyzed by XPS after thermal treatment at 410°C of CsI deposit. Atomic surface composition is detailed in Table I. Cs and O dominates the surface composition whereas the SS elements are detected in low or trace amounts (Mn, Ni, Fe and Cr). Chemical states of these elements are attributed to Cr^{3+} (peak at 275.5 eV, [5]), Fe^{3+} (710.5 eV, [5]), Mn^{3+} (639.4 eV [5]), and Ni^{2+} (855.7 eV [5]). However, the surface is mainly composed of residual Cs compounds. No iodine is detected on the surface, thus iodine is completely revaporized from the coupon.

Table I: XPS surface composition after revaporization at 410°C

Elements	O	Cs	C	Mn	Cr	Ni	Fe
Atomic composition (%)	59.7	24.4	9.0	3.7	0.7	1.3	1.2

Cs $3d_{5/2}$ XPS spectrum presents a single component at 724.0 eV which is attributed to Cs^+ [5]. All I is revaporized during the thermal treatment. Thus, Cs remaining in the residue is not under the CsI form. The focus is then placed on oxygen, the main surface element.

O 1s XPS spectrum features a broad peak (Figure 1) which can be decomposed into two components at 531.4 eV (40%) and 529.4 eV (60%). The higher BE component could be mainly attributed to hydroxide while the lower BE component at oxide [5]. Given the atomic OH^-/Cs ratio close to 1, the presence of CsOH can be assumed. Cs as oxide (Cs_2O, Cs $3d_{5/2}$ BE = 725.2 eV [5]) is not considered due to its instability under air atmosphere [6].

Complementary surface analysis by ToF-SIMS (Bi^+ beam) revealed the significant presence of CsO^+, $CsOH^+$ and $CsOH_2^+$ fragments confirming the nature of CsOH residue on the sample. No fragment of Cs-C-O or Cs-M-O has been found (M corresponding to metal as Mn, Cr, Ni, or Fe).

Thus, according to these complementary surface results, the residue after revaporisation at 410°C is composed of CsOH.

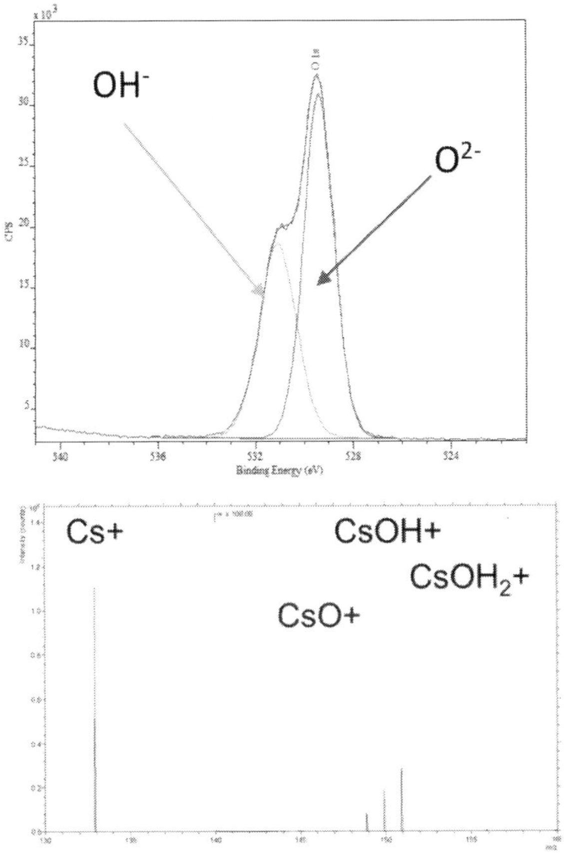

Figure 1: O 1s XPS and ToF SIMS spectra of CsI deposited on SS coupon after revaporization at 410°C in air/steam atmosphere

3.2 Cs speciation after revaporization at 520 °C

Remaining elements on the SS coupon were analyzed by XPS after thermal treatment of CsI at 520°C. Atomic surface composition is detailed in Table II. Cs and O dominate the surface composition. Except for Cr which is detected in low amounts (4.7%), the other substrate elements (Fe, Ni, Mn) are below the detection limit as well as iodine.

Table II: XPS surface composition after revaporization at 520 °C

Elements	O	C	Cs	Cr (+VI)	Cr (+III)
Atomic composition (%)	53.6	30.0	11.7	2.1	2.6

No iodine is detected on the surface, thus iodine is completely revaporized from the coupon.

After revaporisation at a higher temperature (520 °C), Cs is detected in lower amount (11.7%) compared to the revaporisation at 410 °C (24.4%). The Cs $3d_{5/2}$ peak shows one component at 724.0 eV corresponding to Cs^{+}. As hydroxide is also detected at the surface, Cs can be present as CsOH as previously at 410 °C. Moreover, chromium is significantly detected at the surface. Figure 2 shows that Cr $2p_{3/2}$ can be decomposed into two components: one at lower BE at 575.5 eV, attributed to Cr+III [5] and the second at 580.0 eV, attributed to Cr+VI [5]. It was shown in a previous study [7] that the appearance of Cr +VI after CsI revaporisation is due to the formation of caesium chromates as Cs_2CrO_4, as evidenced by Raman spectroscopy.

Thus, species based on Cs-Cr mixed oxide forms (featuring Cr(+VI)) can also be assumed. ToF-SIMS analysis of the sample confirmed this assumption with the detection of $CsCrO^{+}$ (Figure 2) and $Cs_2CrO_2^{+}$ fragments. Based on these data, we can deduce that Cs in the residue is mainly CsOH (90% of total residual Cs)) and also Cs-Cr mixed oxide form (10%) identified as Cs_2CrO_4 This oxide is indeed the main mixed Cs-Cr oxide form reported in the literature in the same conditions [4], [8], [9]).

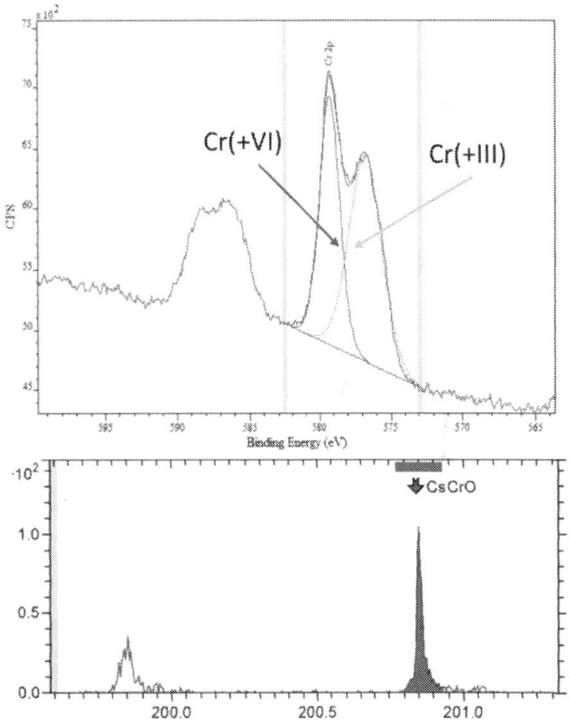

Figure 2: Cr 2p XPS and ToF SIMS spectra of CsI deposited on SS coupon after revaporization at 520°C.

3.3 Iodine Speciation

After revaporisation at 410 and 520°C, no (or trace below detection limit) iodine is detected on the surface by XPS analysis. However, ToF-SIMS is more surface sensitive than XPS, allowing the detection of iodine fragments. The presence of IO^{-}, IOH^{-}, $CsIO^{-}$ and $CsIOH^{-}$ is highlighted in

Figure 3. Cs_2I^+ is also detected. Thus, the remaining iodine is partly oxidized, partly as CsI. In these experimental conditions, ToF-SIMS allows only a qualitative analysis of detected iodine compounds, the distribution of iodine between these different forms cannot be established.

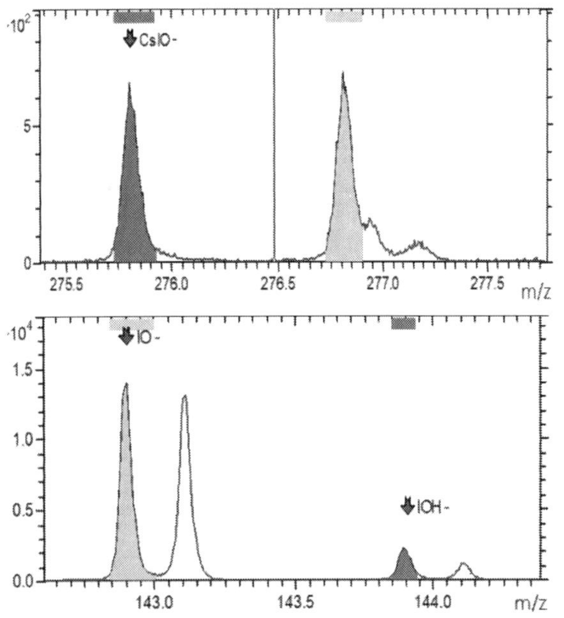

Figure 3: ToF SIMS spectrum of CsI deposit on SS coupon after revaporization at 410°C (with a Bi^+ ion beam and a negative polarity).

4. Conclusion

Analytical experimental work within the ESTER program focuses on the revaporization of synthetic FP deposits, CsI can be considered as a main source of volatile iodine in case of late-phase releases in oxidative conditions. The purpose is to understand revaporization phenomena to depict/characterize a mechanism that took place during the FD severe accident and led to late releases. The revaporization of CsI deposits under different isotherms was studied leading to an increase of the gaseous iodine release (compared to CsI gas) with temperature.

This study focuses on surface analyses of Cs and I residues after revaporization in order to have a better understanding of the revaporization mechanisms. The complementarity of XPS and ToF-SIMS allows the determination of CsOH as the main species in the residue. For the higher temperature, a small part of Cs is under Cs_2CrO_4 form (around 10% of Cs). At the same time, iodine is almost completely revaporized. Traces remaining are under CsI form but also under oxidized form as CsIO and HOI.

These results combined with the increase of gaseous iodine fraction release with temperature may be explained by the two global reactions:

1) The main reaction leading to the formation of CsOH:
$$CsI + \frac{1}{2}O_2 + H_2O \rightarrow CsOH + IOH$$
2) In a smaller proportion leading to Cs_2CrO_4:
$$4CsI + \frac{7}{2}O_2 + 2H_2O + Cr_2O_3 \rightarrow 2Cs_2CrO_4 + 4IOH$$

Acknowledgments

This work was carried out on the surface analysis facility of the Advanced Characterization Platform of the Chevreul Institute, which is thanked for its help in the development of this work through the ARCHI-CM project supported by the "Ministère de l'Enseignement Supérieur de la Recherche et de l'Innovation", the region "Hauts-de-France", the ERDF program of the European Union and the "Métropole Européenne de Lille". This work was supported by the partners (NRC, GRS, SSM, NRA, EDF, VTT, and KINS) of the OECD/ESTER program

References

[1] F. Cousin, *et al.*, *Analysis of fission products behavior and environmental releases during the Fukushima-Daiichi accident by direct and inverse approach at IRSN* in Proceedings, NURETH, 18-23 August 2019, Portland, USA.
[2] N. Girault, *et al.*, Nucl. Technol., **169**, 218–238, (2010). doi: 10.13182/NT10-A9375
[3] J. Kalilainen, *et al.*, Nucl. Eng. Des., **267**, 140–147, (2014) doi: 10.1016/j.nucengdes.2013.11.078
[4] D. Obada *et al.*, J. Hazard. Mater., **409**,124519, (2020), doi: 10.1016/j.jhazmat.2020.124519.
[5] A. Naumkin, *et al.*, NIST Standard Reference Database 20. National Institute of Standards and Technology, (2012) doi: 10.18434/T4T88K.
[6] A. Albu-Yaron et al., Adv. Mater., **18**, 2993-2996, (2006), doi : 10.1002/adma.200600983
[7] A.S. Mamede *et al.*, Appl. Surf. Sci., **369**, 510-519, (2016) doi : 10.1016/j.apsusc.2016.01.185
[8] H. Shibazaki *et al.*, Nucl. Technol., 134, 62–70, (2001) doi: 10.13182/NT01-A3186.
[9] K. Sasaki, *et al.*, *Microstructure analysis for chemical interaction between cesium and SUS 316 steel in fast breeder reactor application* in Proceedings, ICAPP, 24-28 June 2012, Chicago, USA

Characterization of model and real catalysts by APXPS

Virginia Pérez-Dieste[1*]

[1]ALBA Synchrotron Light Source, Carrer de la Llum 2-26, 08290, Cerdanyola del Vallès, Barcelona, Spain

Abstract. In this contribution, I first briefly summarize some of the recent advances relevant for the investigation of heterogeneous catalysis with Ambient Pressure X-ray Photoelectron Spectroscopy (APXPS). In the second part, two examples of the research done at the CIRCE beamline of the synchrotron ALBA are described: CO oxidation on a model curved crystal Pd(111) catalyst and methanol steam reforming on powder bimetallic supported catalysts, PdCu/ monoclinic and cubic zirconia.

1 Introduction

Ambient Pressure XPS typically refers to XPS performed with pressures at the sample of up to a few (~20) mbar, as opposed to standard XPS, for which the analysis chamber needs to be kept in ultra-high vacuum.

Over the last two decades, this technique has earned a place in the repertoire of prominent spectroscopies for in-situ characterization of catalysts' surfaces. A turning point in the development of APXPS was the design by Ogletree et al. in 2002 [1], in which the differential pumping system of the photoelectron analyser, already present in earlier designs, was upgraded with a set of electrostatic lenses to focus the photoelectrons through the small apertures between the pumping stages, significantly increasing the count rate and enabling routine operation at a few mbar.

The distinctive feature of APXPS is the surface and interface sensitivity for operando studies of the electronic structure. The fields of application can be classified according to interface type as those involving solid-gas and solid-vapor interfaces, which comprise heterogeneous catalysis, electrocatalysis, electrodes and batteries in air, wetting, corrosion,... solid-liquid interfaces, including electrified solid-liquid interfaces, and liquid-gas interfaces, with the focus on environmental chemistry.

The information that can be obtained includes the evolution of the oxidation state and chemical environment of surface atoms in reaction, segregation and strong metal support interaction phenomena, changes in the work function, depth profiling and ion distribution in electrified interfaces in operando. It is also possible to monitor the product distribution analysing the gas phase composition above the surface.

A good number of excellent APXPS reviews are available [1-10]. The aim of this contribution is to provide a brief look at some of the most recent advances relevant for heterogeneous catalysis, and to present two examples of APXPS experiments performed at the CIRCE beamline of ALBA.

1.1 Recent advances in APXPS for heterogeneous catalysis

As a relatively young technique, APXPS is evolving and diversifying at a fast pace. One of the explored routes is its simultaneous combination with other in-situ techniques, ensuring the same sample conditions and direct correlation of the information from both sources. Gas product characterization by mass spectrometry, proton transfer-reaction mass spectrometry, or gas chromatography allows to confirm that the reaction is taking place and to monitor activity and selectivity [11].

Ambient pressure Near Edge X-ray Absorption Fine Structure spectroscopy in total and partial electron yield modes is almost by defect available at synchrotron setups, and it gives access to the unoccupied electronic structure of surfaces. Varying the X-ray polarization or its incidence angle, it can be used to determine a preferential orientation of adsorbed gas molecules.

Vibrational spectroscopies like Infrared Reflection Absorption Spectroscopy (IRRAS) and Polarization Modulated-IRRAS are now available at a few synchrotron and lab APXPS setups, and they can be used to ascertain the bonding geometry and adsorption site of gas molecules in model catalysts, or to characterize the active site location and properties by means of titration experiments [12-14].

Recently, APXPS has also been combined with grazing incidence X-ray scattering (GIXS) [15]. Wide and small angle GIXS provide respectively atomic and mesoscopic scale (~200nm) dynamic (μs) structural information that can be correlated with the chemical evolution monitored by APXPS.

In the last few years there has been an impressive progress in the development of time resolution methods for APXPS, key to address the dynamics of catalytic processes.

* Virginia Pérez-Dieste: vperezdieste@cells.es

The hundreds of ms range is already accessible in 3rd and 4th generation synchrotrons directly using the fast, snapshot, mode of MCP/CCD or delay line detectors [8,16]. Essential areas and phenomena like chemical kinetics, detection of short-lived surface intermediates, spectators and active phase identification, segregation or nanoparticle restructuration can be tackled.

To make good use of the time-resolution, a perturbation is needed in order to drive the system out of equilibrium and make the event of interest take place at a larger rate than its opposite. Knudsen et al. have used periodical gas pulses as perturbation, combined with lock-in plus image recognition analysis to achieve a time resolution of 60 ms with very good signal to noise ratio for reversible and repetitive processes [17].

In their method, time resolved data are recorded continuously in snapshot mode during several periodical perturbation cycles and a feature that signals a phase transition on the surface (new species, work function change, etc) is chosen as the locking signal. In this way, the response to each pulse does not need to be exactly the same to be able to add data to accumulate statistics, as the corresponding spectra can be aligned in time using the lock-in signal, detected by the image recognition software. Weak surface and gas phase components simultaneously present can be unveiled and correlated, while preserving the time resolution. The phase transitions during CO oxidation on Pd(100) at 3 mbar and 100 mbar were mapped with down to 60 ms resolution and it was shown that all the surface phases: thick oxide, "Ö5" oxide and CO covered metallic phase, were active to some degree.

Shavorskiy et al. have developed a pump and probe scheme that can reach 20 μs resolution, at the Hippie beamline in Max IV [16]. A fast valve is used to create ms gas pulses in the mbar range. The valve actuator is synchronized with the opening of a shutter on the electron analyzer detector fast camera through a delay generator that allows recording data only at the selected delay. Performing measurements at increasing delays, the time-dependent evolution of the APXPS spectra can be monitored for the whole gas pulse. This scheme also requires that the catalyst response to the gas pulse be reproducible.

The method was applied to the CO oxidation reaction on Pt(111) and it was shown that the higher CO_2 production occurs for $PO_2/PCO=0.9\text{-}2.3$, when the surface contains both chemisorbed oxygen and surface oxides, which was interpreted as indication that both the Langmuir-Hinshelwood and the Mars-van-Krevelen mechanisms intervene in the reaction.

Redekop et al. have implemented the gas pulse perturbation approach at the Species beamline in Max IV using a time counting, delay-line detector. In this proof-of-concept work, emphasis has been put on the careful characterization of the gas transport within the cell during the pulses, obtaining an estimation of the residence time and of the intrinsic reaction rate constant range that can be accessed [18].

Last—and the least in the race against the clock—Favaro et al. have reported achieving 30 ns time resolution with good statistics and energy resolution in a new APXPS endstation using a 3D delay line detector [19].

Another pending task in the APXPS evolution is closing the remaining pressure gap among tenths of mbar operation and industrial catalytic conditions, usually above 1 bar. Since 2019, the 1 bar milestone can be efficiently reached at the Polaris instrument in Dessy, using tender X-rays to increase the photoelectron mean free path and optimizing all the rest of parameters: sample to nozzle distance, nozzle aperture, sample position stability, analyzer transmission, ... [20]. The surface sensitivity (~10Å) can be preserved exploiting grazing incidence geometry (~0.05°-1°). Depth profiling in a large range (~10-150Å) is possible by sweeping the angle, and even larger depths can be reached by increasing the photon energy. For example, it was shown that the Pd(100) surface adopts an island structure during CO_2 oxidation at 300 mbar, with metallic islands growing on top of oxide islands [21].

The Polaris setup uses the gas beam dosing approach, consisting of creating a local high pressure area at the sample position while maintaining the background pressure in high vacuum [22]. This strategy is useful for gas-pulse time-resolved experiments, as it allows a fast exchange of the gas in the high pressure region [17]. The routine operation pressure is currently in the 300-500 mbar range, although 2.5 bar can be reached.

Blomberg et al. showed the potential of this instrument to explore a wide pressure range, obtaining a global picture of the pressure dependence of surface species, products and intermediates, in a Pd(100) surface during CO oxidation with diluted gases [23]. More recently, other reactions of high practical interest, like the Fisher-Tropsch reaction on Fe(110) and the CO_2 hydrogenation on Zn/ZnO/Cu(211) have been addressed [24-25].

2 APXPS for heterogeneous catalysis at ALBA

Model heterogeneous catalysts with usually simple and known structures facilitate experimental reproducibility and admit first-principles calculations closer to the experimental configuration, both things favoring solid understanding of partial processes. On the other hand, more realistic, oxide supported, diluted, multimetallic, structurally defective, multiphase... catalysts preserve most of the complex and dynamic interactions among their components. Next, two examples of APXPS studies, for model and realistic catalysts, both for relatively simple reactions, are summarized.

2.1 CO oxidation in curved Pd(111)

The CO oxidation reaction has been extensively examined by APXPS as a model reaction in model catalysts, in mono and bimetallic supported nanoparticles, in oxides nanorods,... [4]. In this work a Pd(111) curved crystal was used [26]. Curved crystals contain a collection of vicinal surfaces; therefore they represent an intermediate case between single crystals

 https://doi.org/10.1051/epjconf/202227301005

and nanoparticles, and allow studying a variety of crystallographic planes with different atomic coordination sites in a single experiment, in the same experimental conditions.

The c-Pd(111) was exposed to a 1:1 $CO:O_2$, 0.15 mln/min flow at P=0.6 mbar, at 370 K. The temperature was increased up to 570 K, and then decreased to 470 K. The sample was scanned with the 100 x 20 μm^2 beam, probing the different vicinals with an angle step of 2°. At 570 K the whole surface was active, as indicated by the CO_2 production and the absence of adsorbed CO. In Fig. 1, the C1s and O1s spectra for T=485 K during cooling-off are shown.

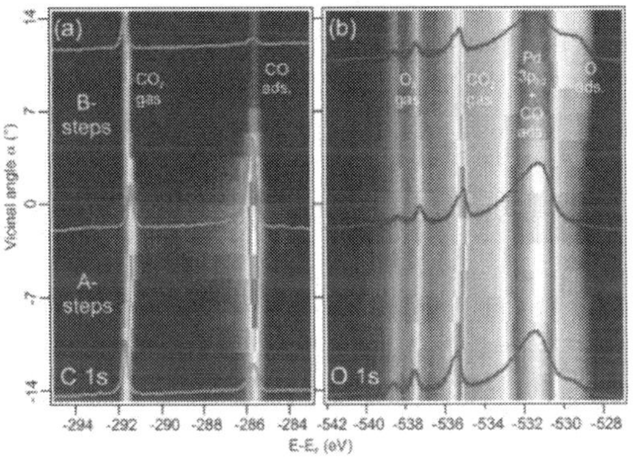

Fig. 1. C1s (a) and O1s (b) intensity plots across the curved surface at 485K, during reaction cooling-off. In (a), the strong variation of the chemisorbed CO indicates that the reaction is poisoned at the sample center, becoming more active towards the A side and fully active at the B edge. Reprinted with permission from [26]. Copyright (2018) American Chemical Society.

The co-adsorbed CO intensity is maximum at the Pd(111) (α=0°) and it decreases with increasing vicinal angle, more markedly towards the B-type (111) steps direction. The O1s intensity exhibits a complementary variation, with an additional feature at BE=529 eV assigned to chemisorbed oxygen and surface and sub-surface oxides. Then, at 485K the catalytic oxidation of CO is switched off at the (111) surface, but it is still ongoing at highly stepped crystal planes and the B-type steps are more active than the A-type steps. At 570 K the oxides are present also in the (111), however at 485 K they only persist in highly stepped vicinals, indicating that they are more difficult to reduce. Their presence could also be favoured by the gas mixture becoming locally more oxidizing when the sample region is CO depleted [23] .

2.2 $PdCu/ZrO_2$ for methanol steam reforming

In this work we investigated Pd-Cu bimetallic catalysts supported on zirconia for the methanol steam reforming (MSR) reaction to produce H_2 [27]. Methanol is liquid at room temperature, what makes it easy to transport, and it is a well suited fuel for H_2 production due to its high H/C ratio and because it does not have C-C bonds, which leads to a lower reforming temperature than for other alcohols .

Cu-based catalysts are the most used for MSR due to their high activity and selectivity to H_2 and their low CO production and cost. However, they have low thermal stability and deactivate easily. To minimize that problem, they are usually supported on oxides and co-catalysts are added. Pd is a good candidate because it presents good thermal stability and relatively low CO selectivity. The metal loading was 20% Cu and 4% Pd.

Two different catalysts with monoclinic and cubic zirconia were prepared and the fixed bed reactor catalytic tests indicated that the catalyst with monoclinic ZrO_2 had higher catalytic activity, higher methanol conversion (90% vs. 40% at 240°C) and lower CO production (5% vs. 20% at 240°C) than the one with cubic zirconia support. The goal of the study was trying to explain the origin of such performance differences.

The X-Ray Diffraction and Scanning Electron Microscopy results revealed that the metal dispersion was better in the monoclinic catalyst, with a mean Cu crystal size ~2 times smaller than in the cubic catalyst.

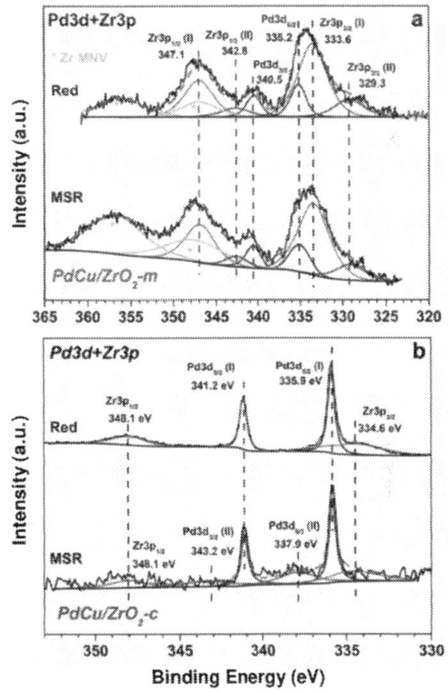

Fig. 2. Pd3d+Zr3p core levels acquired with hv=500eV in reducing and MSR conditions in monoclinic (a) and cubic (b) ZrO_2 supported catalysts. Reproduced from [27]

For the APXPS experiments, the powder samples were pressed into pellets and first pre-treated at 1 mbar H_2 and 300°C for 2 hours to clean and reduce the metals. Then APXPS during the MSR reaction was performed at 1mbar CH_3OH: H_2O 2:3 at T=180 °C
In Fig. 2, the Pd 3d plus Zr 3p spectra in reduction and during the MSR reaction are shown. In the monoclinic catalyst two ZrO_2 components are present both in reduction and in reaction, assigned to different oxidation states, probably Zr^{+4} and Zr^{+3}. On the other hand, in the cubic catalyst the Zr component is very weak and only one component was identified, assigned to Zr^{4+}.

In the cubic catalyst the Pd is in metallic state during reduction and in reaction there is a small oxide component. In the monoclinic catalyst the Pd $3d_{3/2}$ appears at lower binding energy than the metallic Pd, what could be due to PdCu alloying.

The Pd/Zr atomic ratio was calculated from the data and a 15 times lower value was found for the monoclinic catalyst than for the cubic one, which indicates a better dispersion of the metals in the catalyst surface. From the area ratio change, it is also deduced that in the cubic catalyst the Pd segregates towards the surface during MSR, while for the monoclinic catalyst Pd is stable.

Then, it was concluded that the cubic sample presented metal agglomeration, surface oxidation and metal segregation towards the surface under working conditions. In contrast, PdCu alloy is probably present in the monoclinic catalyst, preventing in that case the oxidation under reaction conditions. In addition, the metal distribution is stable during reaction in the monoclinic catalyst and it presents a mixture of oxidation states with vacants that can promote water activation. The good dispersion of the active phase also contributes to the observed enhanced MSR activity.

XPS experiments with realistic catalysts are often hindered by charging of the oxide support. Zirconia, for example, is a large band gap material, ~5 eV, but this experiment was feasible thanks to the relatively large metal loading and to the presence of gas in APXPS, which helps discharging [10]. Other tricks to minimize charging are making the pellets as thin as possible, making the pellets sandwiching a thin Au mesh, or measuring near to the usually metallic sample clamp. Heating and a reducing pre-treatment may also help. Nevertheless, some peak broadening could not be totally excluded in this case, maybe favoured by sample inhomogeneity. Surface spectromicroscopy with moderated spatial resolution has been used to check and correct the effects of an inhomogeneous sample potential, making studies with complex catalysts more accessible [28].

In the current global context, rational optimization of industrial catalytic processes is necessary as part of a sustainable energy roadmap, and it can be affirmed that APXPS is quickly developing in the right directions to contribute to this challenge.

Acknowledgments

I would like to thank all the authors of the two works included in this contribution.

References

1. D.F. Ogletree, H. Bluhm, E.D. Hebenstreit and M. Salmeron, Nucl. Instrum. Methods Phys. Res. A, **601,** 151 (2009)
2. H. Bluhm, J. Electron Spectrosc. Relat. Phenom., **177,** 71 (2010)
3. D. E. Starr, Z. Liu, M. Hävecker, A. Knop-Gericke, and H. Bluhm, Chem. Soc. Rev. **42**, 5833 (2013).
4. J. Schnadt, J. Knudsen, and N. Johansson, J. Phys.: Condens. Matter **32**, 413003 (2020).
5. *"Ambient Pressure Spectroscopy in Complex Chemical Environments"* (A. R. Head, S. Nemšák and B. Eren, ACS Symp. Series, 2021)
6. W. R. Flawell, R.S. Chemistry. (2022) DOI: 10.1039/d2fd00071g
7. S. Zafeiratos, T. Dintzer, D. Teschner et al. J. Catal., **269**, 309 (2010)
8. K. Roy, L. Artiglia and J.A. van Bokhoven *ChemCatChem* **10** 666 (2018)
9. Y. Han, H. Zhang, Y. Yu, and Z. Liu, ACS Catal., **11**, 1464, (2021)
10. R. Dupuy, C. Richter, B. Winter et al. J. Chem. Phys. **154**, 060901 (2021)
11. A.V. Bukhtiyarov, I. P. Prosvirin, M. A. Panafidin et al. Nanomat., **11**, 3292 (2021)
12. A. R. Head, O. Karslioğlu, T. Gerber et al. Surface Science **665,** 51 (2017)
13. K. I. Hadjiivanov, G. N. Vayssilov, Adv. Catal. **47**, 307 (2002)
14. K. Mudiyanselage, D. J. Stacchiola, "In-situ *Characterization of Heterogeneous Catalysts"*, (John Wiley & Sons, Inc., 2013, pp. 209)
15. H. Kersell, P. Chen, H. Martins et al., Rev. Sci. Inst. **92**, 044102 (2021)
16. A. Shavorskiy, G. D'Acunto, V. Boix de la Cruz et al. ACS Appl. Mater. Interfaces, **13**, 47629 (2021)
17. J. Knudsen, T. Gallo, V. Boix et al. Nat. Commun., **12**, 6117 (2021)
18. E. A. Redekop, N. Johansson, E. Kokkonen, et al. Rev. Sci. Instrum., **92**, 044101 (2021)
19. M. Favaro, P.C.J. Clark, M. J. Sear, Surf. Sci. **713,** 121903 (2021)
20. P. Amann, D. Degerman, M.-T. Lee et al. Rev. Sci. Instrum., **90**, 103102 (2019)
21. C.M. Goodwin, M. Shipilin, S. Albertin, J. Phys. Chem. Lett. **12** 18, 4461 (2021)
22. J. Pantförder, S. Pöllmann, J. F. Zhu Rev. Sci. Inst. **76**, 014102 (2005)
23. S. Blomberg, U. Hejral, M. Shipilin, et al., ACS Catal., **11**, 9128 (2021)
24. M. Shipilin, D. Degerman, P. Lomker, ACS Catal. **12**, 7609 (2022)
25. P. Amann; B. Klotzer, D. Degerman, Science, **376**, 6593, 603 (2022)
26. F. Schiller, M. Ilyn, V. Pérez-Dieste, et al., J. Am. Chem. Soc., **140**, 16245 (2018)
27. D. Ruano, B.M. Pabón, C. Azhenha et al. Catalyst **10**, 1005 (2020)
28. P. Bargiela, V. Fernandez, C. Cardinaud et al. Appl. Surf. Sci., **566**, 150728 (2021)

Towards complete band structure of microscopic MoS_2 flakes.

Sergey Babenkov[1], Marie Froidevaux[2], Peng Ye[1], Ludovic Tortech[1], Yannick Dappe[1], Willem Boutu[1], Nickolas Barrett[1], and Hamed Merdji[2]

1 Université Paris-Saclay, CEA, CNRS, 91191 Gif-sur-Yvette, France
2 LOA, ENSTA Paris, CNRS, Ecole Polytechnique, 91762 Palaiseau, France

Abstract. The occupied and unoccupied electronic states of MoS_2 monolayer isolated flake were studied using laboratory based photoemission electron microscope (PEEM) nanoESCA equipped with He-I photon source. PEEM real-space imaging allowed selecting the high quality flake. Altogether, the data will allow accurately recovering the band structures of MoS_2. The band structures will be used in future pump-probe experiments to explore the dynamics of electrons in the conduction band and photo-induced multi-topological states using trefoil polarization.

Introduction

Since the discovery of graphene in 2004, the interest of solid-state physics and the material science community in graphene-like two-dimensional (2D) materials has exploded. During the past decade, various methods have been developed to synthesize and manipulate them. Transition-metal dichalcogenides (TMDC) (MX_2 layers with 2H symmetry where M = Mo, W; X = S, Se, Te) occupy a special place in the class of 2D-materials thanks to their fascinating properties [1, 2, 3]: direct band gaps at monolayer thickness, both chemically inert and stable surfaces at room temperature and topological states.

Recently, it has been theoretically shown that a topological phase transition can be stimulated entirely with shaped light fields, such as trefoil polarization states, in conventional hexagonal materials [4]. The idea is based on nonlinear interaction of monolayer MoS_2 and the bicircular field (trefoil polarization). Such light fields with trefoil symmetry [5] and complex orbital angular momenta [6] have already been generated using higher laser harmonics. Trefoil polarization can be created combining ω and 2ω photons with circular polarization of opposite helicity. In the experiment, the orientation of the trefoil with respect to MoS_2 structure can be controlled either by phase delay φ between two pulses or by sample rotation. For example, when $\varphi = -\pi/2$, both sublattices are addressed equally. As a result, the K and K' valleys are nearly equally excited. When $\varphi = 0$, the cycle-averaged band structure shows a strong valley asymmetry. This reflects in the valley populations, which show strong contrast. The situation is reversed for $\varphi = \pi$, where the electron population switches to the opposite valley. The overall helicity of the field also contributes to the valley asymmetry, but its effect is relatively weak. It manifests in small valley polarization for $\varphi = \pm\pi/2$, and in that $\varphi = 0$ and $\varphi = \pi$ are not exact opposites.

This opens a route towards a new device based on electronic topology and ultrashort light pulses. In particular, by orienting the trefoil symmetry it is possible to populate alternatively K and K' valleys in the conduction band and measure their dynamics using pump-probe experiments. This is one of the goals of our H2020 project "OPTOlogic" [7].

The first question is therefore to measure the band structure at the surface of a single flake. The possibility to obtain the information about unoccupied states was predicted [8,9] and recently explored using laboratory based equipment [10]. In practice, the k-space intensity maps taken in the energy range of secondary electrons, secondary electron emission microscopy (k-SEEM), will consist of the local intensity minima and maxima, which indicate the unoccupied bands. The available energy range depends on the electron effinity of the given sample. The lower electron affinity is, the more closer vacuum level is to conduction band minimum (or even vacuum level lies below conduction band minimum), the large is available energy region in *k*-SEEM.

With this idea in mind, in this work we studied the occupied and unoccupied states of isolated MoS_2 monolayer flakes on top of SiO_2/Si wafer by means of XPS, k-PEEM and k-SEEM.

Experimental details

In this study, we were using commercially available CVD-grown monolayer 2H-MoS2 flakes purchased from 6carbon technology [11]. The flakes are grown on top of a 300nm-SiO2/Si wafer (source substrate) which makes the photoemission experiments impossible due to the charging of the SiO_2 surface. One way to overcome this problem is to transfer the flakes from source substrate to Si wafer with thin (2-3 nm) SiO_2 layer (target substrate). It has 2 advantages: (i) the layer of the

SiO_2 is thick enough to supress the band features of the Si crystal allowing the photoemission experiment and (ii) thin enough to avoid possible charging effects using X-ray photoemission spectroscopy (XPS), Figure 1. XPS spectra were recorded at 3 different emission angles: normal (90°), intermediate (55°), grazing (20°). No detectable shifts due to charging of the surface was observed, while the thickness of the SiO_2 was estimated to be 2-3 nm.

In order to transfer flakes from source to target substrate, we used surface energy-assisted transfer which is described in detail elsewhere [12]. In short, the source wafer was covered by the thin membrane of Polystyrene using polystyrene/toluene solution and spin-coating (3500RPM , 60´´). Later the polystyrene membrane is dried on the hot plate (80°C, ~15´). Using deionized water we detached the membrane and flakes from the wafer. The floating membrane was caught by a target wafer. The residual water between membrane and wafer surface was removed by a paper towel following by gentle heating on the hot plate (40°C, ~20´). The polystyrene membrane was dissolved in toluene bathes in 3 steps (~ 20´ each). Finally, the wafer was splashed with ethanol and dried by purified nitrogen.

The residual contamination was desorbed by annealing at 450°C during 30 min in nanoESCA preparation chamber. It allowed considerably improve the quality of the k-PEEM images. The sample was *in-situ* moved to the analytical chamber for PEEM measurements.

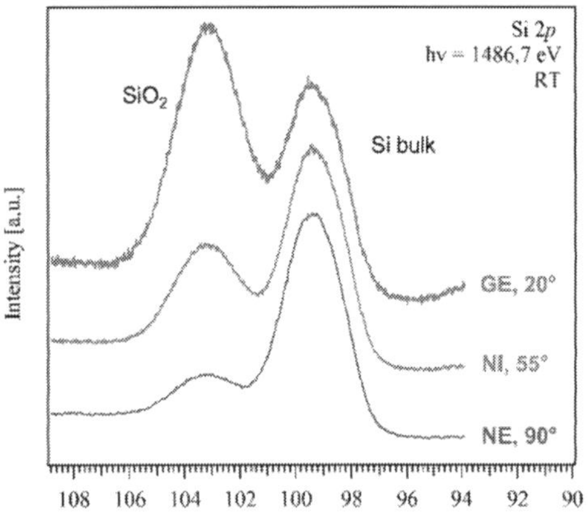

Figure 1. XPS Si 2p spectra taken at different electron emission angles from normal (90°) to grazing (~20°).

The nanoESCA analytical chamber is equipped with Hg (~ 4.9 eV) and He-I (21.2 eV) photon sources allowing best real-space and k-space resolution. The sample manipulator in nanoESCA analytical chamber allowing reproducible precise manipulation with sample and bring the region of interest to field of view even after transfers of sample to the preparation chamber and back to analytical chamber. Therefore, it is possible to study exactly the same MoS_2 flake after sample annealing and/or deposition of thin films.

Results and discussion

Figure 2 presents large field of view (a) and detailed (b) real-space images of the MoS_2 flakes taken using Hg (hv~ 4.9 eV) photon source. It is clear that the flakes are distributed randomly in terms of size, shape and orientation over the sample surface. Therefore, in order to select single isolated flake of best quality for k-PEEM measurements the real-space microscopic capabilities of nanoESCA are essential. The region selected for k-PEEM measurements is highlighted on figure 2 (b) by yellow lines.

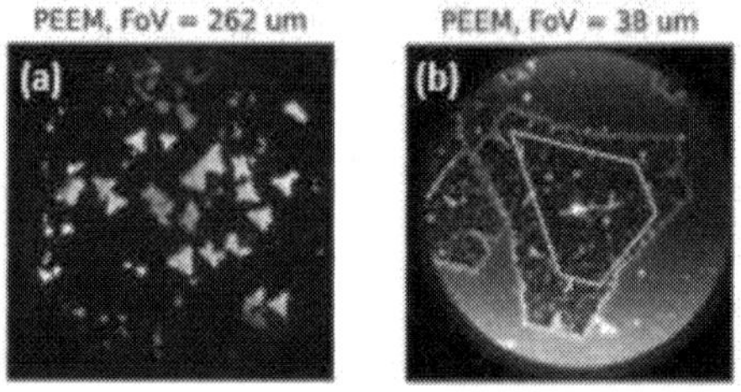

Figure 2 Large field of view (a) and detailed (b) real space PEEM images taken using Hg photon source (~4.9eV). The region of the flake selected by iris for k-PEEM is highlighted by yellow lines.

Thanks to the thin SiO_2 chemical oxide, there are no detectable image and spectra distortions caused by charging under photoemission. Consequently, the polystyrene-based chemical transfer allows creating $MoS_2/SiO_2/Si$ sample of sufficiently good quality for photoemission experiments avoiding intermediate conduction buffer layer, as was proposed by S.W. Jung et al. [13].

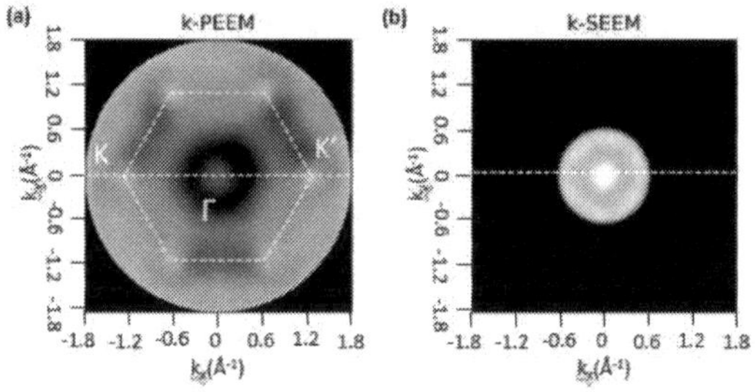

Figure 3 *2nd derivative of constant energy cuts I(E; kx, ky) taken on single MoS2 flake: (a) valence band structure taken at 1.4 eV binding energy and (b) the structure of unoccupied states in the second electron emission tail at 15.5 eV binding energy. The data were acquired using He-I emission (hv = 21.2 eV) at 30K sample temperature.*

Figure 3 presents the second derivative of *k*-photoemission electron microscopy (k-PEEM) momentum sections $I(E; k_x, k_y)$ in the regions close to Fermi level (a) and in the secondary electron tail (b) taken on a single MoS_2 flake. On panel (a) the valence band structure at 1.4 eV binding energy is presented. The high symmetry points K, K´found themselves at the vertices of hexagon and Γ in the center of hexagon are highlighted. It is clear that the intensity of the signal in panel (b) is strongly modulated.

Fig. 4 presents the E-*k* cuts through K-K´direction of the data stacks along the straight dashed lines in Figure 3(a) for k-PEEM region and Figure 3 (b) for k-SEEM region. Black dashed lines in figure 4 indicate on the slice where the data for figure 3 were taken. In case of k-SEEM the accessible region of the energy and $k_{\|}$ is limited by electron affinity and free-electron dispersion parabola defined by the relation $E = \frac{\hbar^2 k_{\|}^2}{2m_e}$. It is highlighted by yellow dashed line. One way to get information about the unoccupied states closer to the conduction band minimum is to lower the effective work function (WF) of the sample, for example by deposition of Cs thin film. The expected available region is highlighted by blue dashed line. The figure 4 presents the calculated occupied and unoccupied states. First-principles calculations have been performed using a DFT localized orbital molecular dynamic technique (FIREBALL) [14-17]. In this study, we have considered a standard (1 x 1) unit cell of MoS_2. A set of 300 special k points along the G-K-M path has been used for the band structure calculations.

In case of occupied states, the k-PEEM data are in good agreement with theory. However, there is a clear difference in theory and experimental data on unoccupied states. One of the possible reasons is that the available kinetic energy range given by the He I photon beam is too short (~17eV). The occupied and unoccupied states could overlap. In order to expand the kinetic energy range and separate the states it is necessary to use a photon beam of higher energy, for example at the synchrotron (hv~100-200eV). In this case, the secondary emission tail will appear at much higher kinetic energies that allows large kinetic energy gap between the occupied states near fermi level and unoccupied states in secondary emission tail.

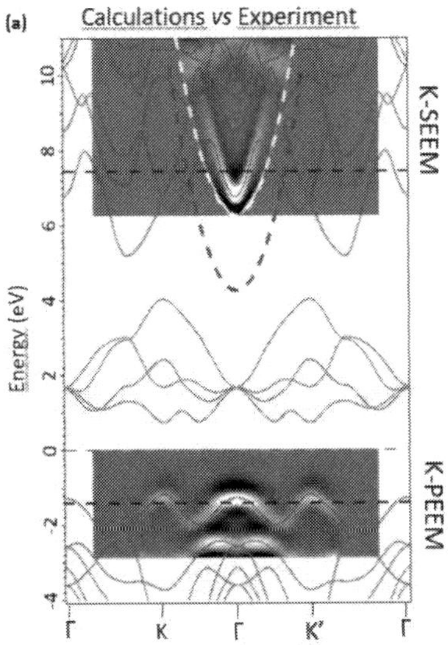

Figure 4. Calculated occupied and unoccupied electronic states and corresponding experimental k-PEEM. E-k cuts through K-Γ-K' as shown by horizontal, white dashed lines in Figure 3 (a, b).

Conclusions

The presented results shows that the band structure of MoS_2 isolated monolayer flakes on 2-3nm SiO_2/Si wafer can be studied by means of photoemission experiment. No harmful surface charging effects were detected. The possibility to study the unoccupied states was limited by the available kinetic energy range provided by He I emission. This could be overcome by using photons of sufficiently higher energy, for example using synchrotron radiation. In addition, it is possible to lower work function by deposition of Cs thin film on the surface of the sample.

Acknowledgements

We want to acknowledge Dr. Vincent Derycke and Nathan Ullberg (CEA-Saclay, IRAMIS) for help with sample preparation and fruitful discussion of the results. We want to acknowledge support from OPTOlogic European project Nr EUR830OPTOLO.

References

1. C. Jin, E.Y. Ma, O. Karni, E.C. Regan, F. Wang and T.F. Heinz., Nature Nanotechnology **13**, 994–1003 (2018)

2. M. Chhowalla, H.S. Shin, G. Eda, L.-J. Li, K. P. Loh and H. Zhang, Nature Chemistry **5**, 263–275 (2013)

3. B.A. Bernevig, T.L. Hughesand, and S.-C. Zhang, Science **314**, 1757 (2006)

4. R.E.F. Silva, Á. Jiménez-Galán, B. Amorim, O. Smirnova and M. Ivanov, Nature Photonics **13**, 849–854 (2019)

5. A. Fleischer, O. Kfir, T. Diskin, P. Sidorenko and O. Cohen, Nature Photon **8**, 543–549 (2014)

6. L. Rego, K.M. Dorney, N.J. Brooks, Q.L. Nguyen, C.-T. Liao, J.S. Román, D.E. Couch, A. Liu, E. Pisanty, M. Lewenstein, L. Plaja, H.C. Kapteyn, M.M. Murnane, and C. Hernández-García., Science **364** (6447), 1253 (2019)

7. https://cordis.europa.eu/project/id/899794

8. R. Feder and J. B. Pendry, Solid State Commun. 26, 519 (1978)

9. M. Bovet, V. N. Strocov, F. Clerc, C. Koitzsch, D. Naumović, and P. Aebi 1, Phys. Rev. Lett. 93, 107601 (2004)

10. G. Wan, A. Croot, N.A. Fox, M.Cattelan, Adv. Funct. Mater. **2021**, 31, 2007319

11. http://www.6carbon.com/index-en.php

12. A. Gurarslan, Y. Yu, L. Su, Y. Yu, F. Suarez, S. Yao, Y. Zhu, M. Ozturk, Y. Zhang, and L. Cao, ACS Nano 2014, 8, 11, 11522–11528

13. S.W. Jung, S. Pak, S. Lee, S. Reimers, S. Mukherjee, P. Dudin, T.K. Kim, M. Cattelan, N. Fox, S.S. Dhesi, C. Cacho, and S.N. Cha, Applied Surface Science **532** (2020) 147390

14. J. Lewis, et al., Phys. Rev. B 64 (2001) 195103.

15. J.P. Lewis, et al., Phys. Status Solidi 248 (2011) 1989-2007.

16. P. Jelínek, H. Wang, J. Lewis, O. Sankey, J. Ortega, Phys. Rev. B 71 (2005) 235101.

17. O.F. Sankey, D.J. Niklewski, Phys. Rev. B 40 (1989) 3979-3995.

Methodological developments to expose and analyse buried interfaces in lithium solid-state batteries using *ex situ, in situ* and *operando* cycling

Julien Morey[1] *, *Jean-Bernard Ledeuil*[1], *Lénaïc Madec*[1,2*], and *Hervé Martinez*[1,2,3*]

[1] Universite de Pau et des Pays de l'Adour, E2S UPPA, CNRS, IPREM, Pau, France
[2] Réseau sur le Stockage Electrochimique de l'Energie, CNRS FR3459, Amiens, France
[3] Ecole Centrale Casablanca, Centre de Recherche Systèmes Complexes et Interactions, Bouskoura (Maroc)

Abstract. Lithium solid-state batteries (SSBs) are a promising technology for electrochemical energy storage systems. So far, the performance of SSBs are mainly governed by the electro-chemo-mechanical properties of the diverse solid/solid interfaces and their evolution upon cycling. However, as these interfaces are buried in the battery stack, their comprehensive understanding remains a challenge. Here, we thus provide some advances in methodological developments for *ex situ, in situ* and *operando* cycling/analysis of these buried interfaces. It is showed that noble gaz ion milling at liquid nitrogen temperature is a suitable and reproducible method to prepare cross-section without any chemical/physical change even for polymer-based SSBs. In addition, innovative *operando* cycling using Auger analysis was proposed for the first time on a model Li/Li_6PS_5Cl stack. The interest of this approach is to be able to proceed without a dedicated electrochemical cell and to use the fully adjustable electron beam of the auger to create a surface potential difference followed by lithium migration then SEI (Solid Electrolyte Interface) formation and Li plating. Overall, this work should greatly benefits to all researchers working on buried interfaces study in lithium solid-state batteries.

1 Introduction

Lithium solid-state batteries (SSBs) are currently recognized as the most promising technology for the next generation of Li rechargeable batteries [1]–[3]In these systems, the liquid electrolyte of common Li ion batteries is replaced by a solid electrolyte with the dual role of Li ion conductor and separator between the electrodes. Electrochemical performance of SSBs are mainly governed by (i) ionic and electronic properties of the solid electrolyte and (ii) the chemical and structural properties of the diverse solid/solid interfaces and their stability. Thus, improvement of SSBs electrochemical performance will be achieved through a comprehensive understanding of the interphases electro-chemo-mechanical properties upon cycling. [4], [5]

As illustrated in **Fig. 1**, a significant number of interfaces are identified in SSBs. For instance, the Li/solid electrolyte (SE), positive active material/SE, conductive additive/SE and current collector (CC)/SE are the main interfaces in SSBs. However, the rigorous analysis of all interfaces in SSBs is not trivial as they are buried inside the battery stack, *i.e.* two adjacent materials cannot be easily separated to probe the interphase. Moreover, separation of the two materials can deteriorate the chemical and structural properties of the interphase, which would lead in a loss of relevant information. Suitable and reproducible approaches are thus mandatory to reveal and analyse buried interfaces in SSBs. **Fig. 1** also illustrates the *ex situ, in situ* and *operando* approaches used to analyse buried interfaces of SSBs. Briefly, *ex situ* analysis is the most widely used approach and consists in electrochemically cycling the SSB cell, then disassembled it w/wo specific preparation to reveal the interfaces prior to the analysis. To avoid sample pollution and possible interfaces degradation induced by the cell disassembling and sample preparation, *in situ* and *operando* approaches are very attractive. During *in situ*, the SSB cell is sequentially cycled and analysed inside an analysis chamber while during *operando*, the cycling is performed at the same time of the analysis (**Fig. 1**). Relevant examples of *operando* XPS studies on SSBs can be found in references [6]–[13].

In this paper, we will provide a model approach we recently developed to reveal and analyse buried interfaces in lithium solid-state batteries using *ex situ, in situ* and *operando* cycling.

* Corresponding author: julien.morey@univ-pau.fr , lenaic.madec@univ-pau.fr , herve.martinez@univ-pau.fr

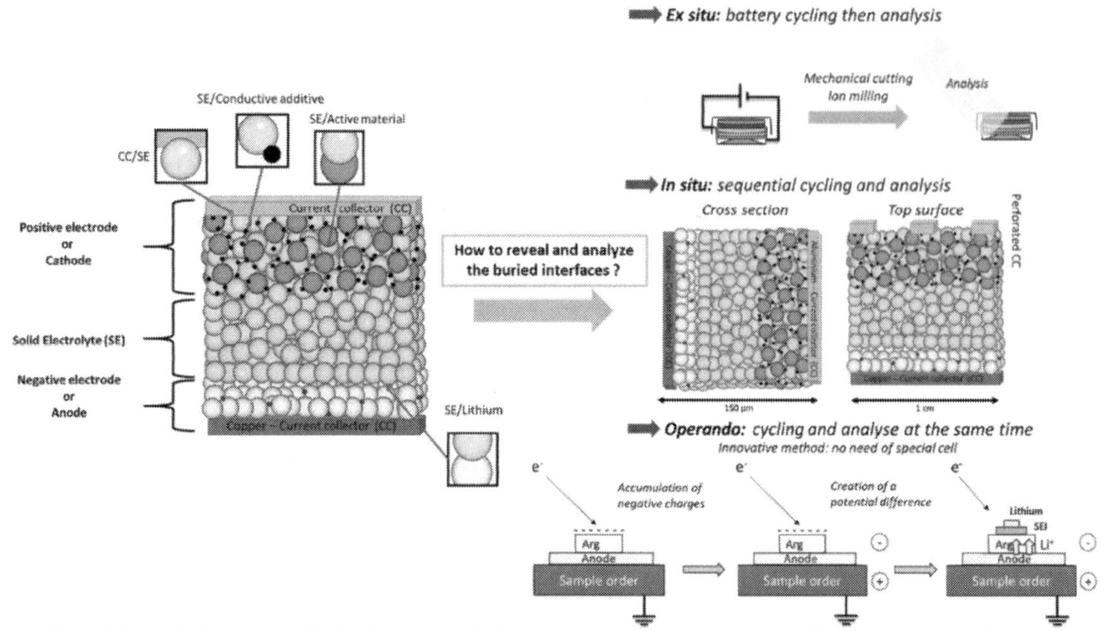

Fig. 1. Schematic illustration of buried interfaces in SSBs and the *ex situ*, *in situ* and *operando* approaches to analyse them.

2 *Ex situ* analysis of buried interfaces in SSBs

To analyse *ex situ* the buried interfaces of SSBs using surface analytical techniques, it is mandatory to reveal these interfaces in a first step. However, numerous literature reports rely on the manual ability of a researcher to separate mechanically the interfaces [5], [14]–[16]. Here, we will show that even for polymer-based SSBs, the use of Ar^+ ion milling (*i.e.* a cross-polisher, **Fig. 2a**) is a suitable method to reveal buried interfaces. In this study, the common poly(ethylene oxide) (PEO) with lithium bis(trifluoromethanesulfonyl)imide (LiTFSI) was used as polymer electrolyte. Commercial graphite and NMC532 electrodes (MTI corp) were used and filled by the PEO-LiTFSI using anhydrous acetonitrile followed by its evaporation at 70°C. Preparing cross-section of such NMC/PEO-LiTFSI/graphite stack is thus a challenge. Indeed, polymers are expected to flow and/or to delaminate and/or to be simply damage by the constraints produced by the Ar^+ beam such as the heat. Therefore, to obtain a perfectly flat cross-sectioned surface without any physical and chemical change, different parameters such as the ion beam voltage/current and working temperature were adjusted by iterative tests. Finally, an Ar^+ beam of 8 kV and 3.5 mA as anode current was used and the temperature was fixed at 133°K using liquid nitrogen cooling in a SEMPREP2 from TECHNOORG LINDA. **Fig. 2b** shows the SEM image of the obtained cross-section for the NMC/PEO-LiTFSI/graphite stack after 6h. As observed, no delamination or flowing of the polymer occurred. To further check that no chemical damage was induced during ion milling, surface analysis was carried out. Auger spectroscopy (AES) and X-ray spectroscopy (XPS) showed that no major difference was observed between the analysis of the cross-section and references (*i.e.* starting) materials (**Fig. 2c** and **2e**). In details, AES analysis showed that no unexpected transition appeared, and the shape line matched perfectly with the reference. Regarding the XPS, no major change was observed on the Mn 2p and Ni 2p except some peaks broadening, characteristic of a less well-defined local chemical environment following the creation of the cross-section or the possible reduction of the transition metals despite that it is not expected for such material in a lithiated environment. This is supported by the fact that no LiF formation was observed from the LiTFSI salt degradation while it would have been expected using Ar^+ milling. Furthermore, characteristic signals from the fluorine, Auger KLL transition and XPS peaks overlapping with the Ni 2p peaks could be identified. Note that the possible degradation was also investigated by ToF-SIMS, which confirmed the XPS and AES results (not shown here). Finally, scanning Auger microscopy (SAM), *i.e.* elemental 2D nano distribution images showed that PEO-LiTFSI electrolyte filled the entire electrodes thickness as deduced from the sulphur mapping attributed to the LiTFSI salt (**Fig. 2d**). Overall, AES, XPS and SAM showed that the ion milling performed at the liquid nitrogen temperature did not chemically/physically damage the revealed surface as no significant change was observed. Thus, this cross-section preparation method combined with *ex situ* cycling can easily be implemented to investigate the interfaces over cycling.

3 *In situ* / *Operando* analysis of buried interfaces in SSBs

3.1 *In situ* cycling

As discussed in the introduction, *in situ* cycling refers to an alternation between electrochemical cycling and XPS ; (ii) from the top surface, using a perforated current collector is a good alternative to overcome possible cross-section limitations. However, interfaces of only one electrode can be probed at once.

analysis. To do so, dedicated cells need to be design according to the analytical machine used. Importantly, two different configurations can be used (**Fig. 1**): (i) using cross-section, all the interfaces are directly exposed but the spatial resolution of the technique may limit the analysis of the Li/SE interfaces, especially in the case of

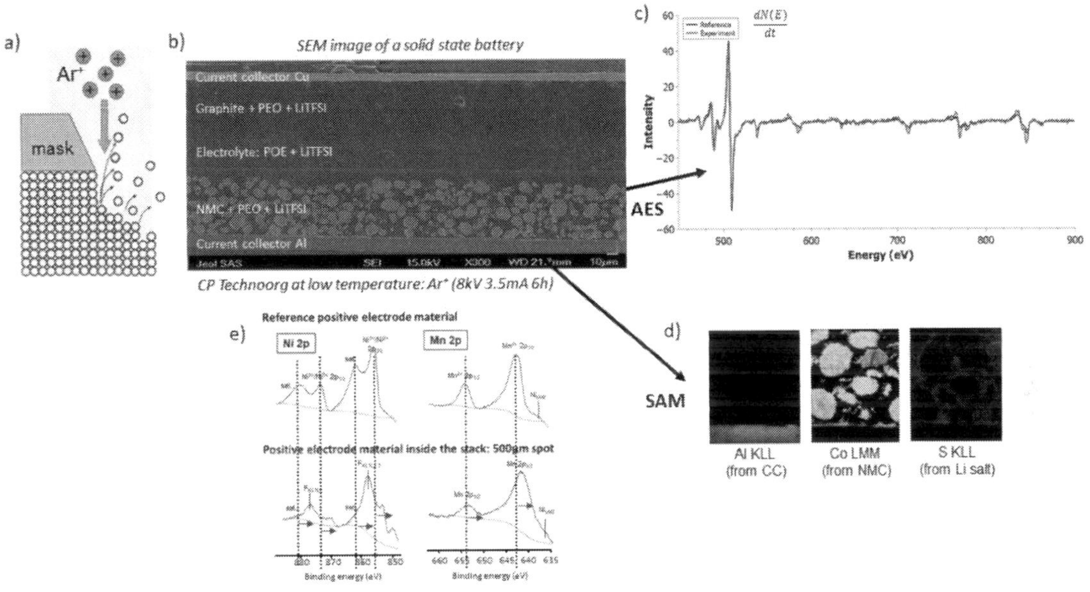

Fig. 2. a) Schematic representation of the ion milling (*i.e.* cross-polishing) principle, b) SEM image of a NMC/PEO-LiTFSI/graphite cross-section obtained after 6h at 8 kV, 3.5 mA and 133°K, c) comparison of AES spectra for the reference NMC electrode and the cross-section, d) SAM (elemental 2D nano distribution) images of the cross-section, e) comparison between XPS spectra from the reference NMC electrode and the cross-section.

3.2 *Operando* cycling

The aim of *operando* cycling is to follow directly the evolution of the materials and interfaces without the minimum sample preparation and concomitant possible alteration. Ultimately, it is then possible to eliminate the use of a dedicated electrochemical cell by using instead the compensation charge gun working in an only electron flooding mode for XPS and ToF-SIMS analysis or the primary electron beam for SEM/Auger analysis (**Fig. 1**) [17]. In that case, the main drawback remains the lack of pressure and voltage control, especially for inorganic SSBs. The most common use and interest of this approach is the study of the Li/SE interfaces formation that cannot be access by *ex situ* or *in situ* analysis. [18]–[20]Typically, a Li/solid electrolyte stack is positioned directly on a sample holder (being grounded) and the electron flow applied on top of the solid electrolyte will create a potential difference by charges build up which will induced Li^+ migration from the lithium electrode (**Fig. 1**). SEI formation and Li plating will then occur. Moreover, the UV gun, often present in XPS equipment, can be used to reversibly strip the Li by removing valence band electrons. Note that is also possible to use a Li/solid electrolyte/positive electrode stack or any other desired configuration and thus charge/discharge the whole assembly by combining flood and UV guns in XPS analysis.

In this work, this innovative *operando* cycling method was performed for the first time using Auger analysis (JEOL JAMP9500F) on a model system composed of a lithium/argyrodite (Li_6PS_5Cl, noted Arg thereafter). **Fig. 3a** shows that after applying a current beam of 5.10^{-8} A, 10 KeV (0.8 mA/cm²) for 2h using a large area scan of 30 x 20 µm, the surface of the Arg became covered by some new compound(s) with a secondary electron black contrast. Auger spectra showed the decrease of the P LVV, S LVV and Cl LVV transition peaks from the Arg and the increase of the Li transition peak (**Fig. 3b**), indicating possible Li plating. This latter was confirmed by the peak shape and position that corresponded perfectly to the Li KVV transition of a pure Li metal reference available in the JEOL database (not shown here).

Again, by comparison with JEOL references, another component transition appeared during the experiment and is attributed to the reduction of Arg into Li_2S (not shown here). The Li_2S area corresponds to the secondary electron white constrast area surrounding the Li metal plating in the SEM image (**Fig. 3b**). Note that Li_2S was also found to form first before Li plating (not shown here). The SAM image overlay of Li_{metal}(KVV), Li_{Li2S}(KVV) and S_{Arg}(LVV) clearly illustrates the Li_2S formation and Li plating (**Fig. 3c**). Due to the shape of the Li(KVV) analysis and the JEOL references a differentiation between Li metal and oxidized Li can be made. Note that these results were confirmed by XPS analysis using the flood gun as electron source in a pair sample (not shown here). Overall, this experiment demonstrates that *operando* cycling in Auger analysis is a very powerful method to study the solid electrolyte reactivity in SSBs. Indeed, using Auger, the electron beam parameters (*i.e.* the current/cm^2 corresponding to the cycling rate) are fully adjustable compared to flood guns in XPS and ToF-SIMS. Thus, *operando* Auger analysis needs to be further develop in the future.

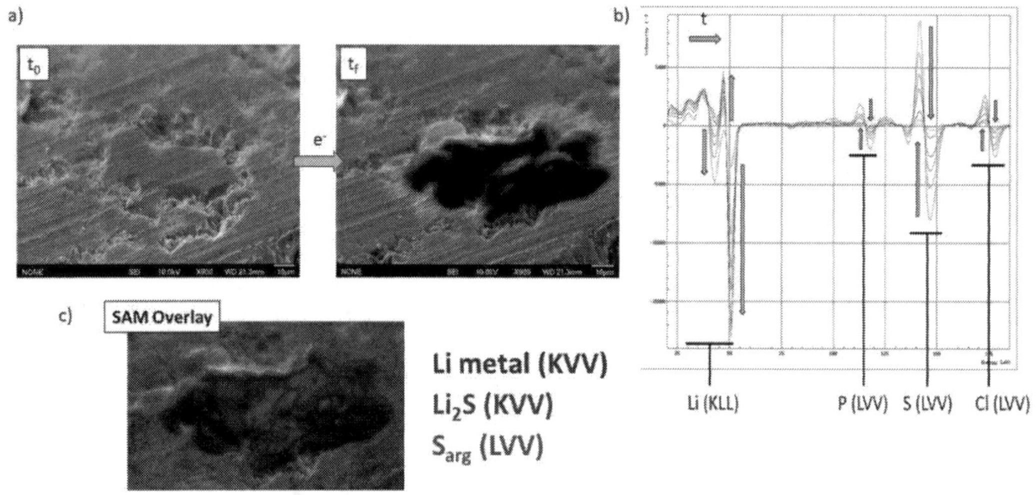

Fig. 3. a) SEM image of Arg deposited on Li metal before and after application of an electron beam using an area scan mode for 2h, b) corresponding AES spectra of the area scan recorded during the operando cycling of the Li/Arg stack c) SAM image overlay of Li_{metal}(KVV), Li_{Li2S}(KVV) and S_{Arg}(LVV) after 2h of acquisition.

4 Conclusions

This work provided latest advances in methodological developments for *ex situ*, *in situ* and *operando* cycling/analysis of buried interfaces in lithium solid-state batteries (SSBs).

Ion milling at liquid nitrogen temperature was proved highly efficient to prepare clean and perfectly flat cross-section of a NMC/PEO-LiTFSI/graphite without any chemical/physical change as observed by XPS, AES, SAM and ToF-SIMS analysis. This result thus highlights that ion milling, with appropriate and controlled parameters, should be preferred to manual cutting as a suitable and reproducible method to reveal buried interfaces even for polymer-based SSBs.

Regarding *in situ* cycling, it is believed that top surface analysis using a perforated current collector should be preferred for XPS while cross-section analysis are more appropriate for high lateral resolution techniques as ToF-SIMS and Auger.

Finally, an innovative *operando* cycling using Auger analysis was proposed for the first time. Using a model Li/Li_6PS_5Cl stack, the SEI formation (Li_2S) and Li plating were observed by applying the Auger electron beam in area scan mode on top of this stack. In other words, that the charges build up which leads to the creation of a potential difference thenthe migration of lithium through the Li_6PS_5Cl solid electrolyte.

Overall, this work should greatly benefits to all researchers working on buried interfaces study in lithium solid-state batteries.

References

[1] R. Chen, Q. Li, X. Yu, L. Chen, and H. Li, "Approaching Practically Accessible Solid-State Batteries: Stability Issues Related to Solid Electrolytes and Interfaces," *Chem. Rev.*, vol. 120, no. 14, pp. 6820–6877, 2020.

[2] J. Janek and W. G. Zeier, "A solid future for battery development," *Nat. Energy*, vol. 1, no. 9, 2016.

[3] Y. Ding, Z. P. Cano, A. Yu, J. Lu, and Z. Chen, "Automotive Li-Ion Batteries: Current Status and Future Perspectives," *Electrochem. Energy Rev.*, vol. 2, no. 1, pp. 1–28, 2019.

[4] A. Banerjee, X. Wang, C. Fang, E. A. Wu, and Y. S. Meng, "Interfaces and Interphases in All-Solid-State Batteries with Inorganic Solid Electrolytes," *Chem. Rev.*, vol. 120, no. 14, pp. 6878–6933, 2020.

[5] I. López, J. Morey, J. B. Ledeuil, L. Madec, and H. Martinez, "A critical discussion on the analysis of buried interfaces in Li solid-state batteries.: Ex situ and in situ / operando studies," *J. Mater. Chem. A*, vol. 9, no. 45, pp. 25341–25368, Dec. 2021.

[6] X. Wu, C. Villevieille, P. Novák, and M. El Kazzi, "Monitoring the chemical and electronic properties of electrolyte-electrode interfaces in all-solid-state batteries using: Operando X-ray photoelectron spectroscopy," *Phys. Chem. Chem. Phys.*, vol. 20, no. 16, pp. 11123–11129, 2018.

[7] X. Wu, C. Villevieille, P. Novák, and M. El Kazzi, "Insights into the chemical and electronic interface evolution of Li4Ti5O12 cycled in Li2S-P2S5 enabled by: Operando X-ray photoelectron spectroscopy," *J. Mater. Chem. A*, vol. 8, no. 10, pp. 5138–5146, 2020.

[8] M. Mirolo, X. Wu, C. A. F. Vaz, P. Novák, and M. El Kazzi, "Unveiling the Complex Redox Reactions of SnO2in Li-Ion Batteries Using Operando X-ray Photoelectron Spectroscopy and in Situ X-ray Absorption Spectroscopy," *ACS Appl. Mater. Interfaces*, vol. 13, no. 2, pp. 2547–2557, 2021.

[9] R. Koerver *et al.*, "Redox-active cathode interphases in solid-state batteries," *J. Mater. Chem. A*, vol. 5, no. 43, pp. 22750–22760, 2017.

[10] Y. C. Lu *et al.*, "In situ ambient pressure X-ray photoelectron spectroscopy studies of lithium-oxygen redox reactions," *Sci. Rep.*, vol. 2, pp. 1–6, 2012.

[11] M. Favaro *et al.*, "Unravelling the electrochemical double layer by direct probing of the solid/liquid interface," *Nat. Commun.*, vol. 7, no. May, pp. 1–8, 2016.

[12] X. Wu, M. El Kazzi, and C. Villevieille, "Surface and morphological investigation of the electrode/electrolyte properties in an all-solid-state battery using a Li2S-P2S5 solid electrolyte," *J. Electroceramics*, vol. 38, no. 2–4, pp. 207–214, 2017.

[13] D. Weingarth, A. Foelske-Schmitz, A. Wokaun, and R. Kötz, "In situ electrochemical XPS study of the Pt/[EMIM][BF4] system," *Electrochem. commun.*, vol. 13, no. 6, pp. 619–622, 2011.

[14] J. Liang *et al.*, "Stabilizing and understanding the interface between nickel-rich cathode and PEO-based electrolyte by lithium niobium oxide coating for high-performance all-solid-state batteries," *Nano Energy*, vol. 78, no. August, p. 105107, 2020.

[15] R. Schlenker *et al.*, "Understanding the Lifetime of Battery Cells Based on Solid-State Li6PS5Cl Electrolyte Paired with Lithium Metal Electrode," *ACS Appl. Mater. Interfaces*, vol. 12, no. 17, pp. 20012–20025, 2020.

[16] L. Wang, D. Liu, T. Huang, Z. Geng, and A. Yu, "Reducing interfacial resistance of a Li1.5Al0.5Ge1.5(PO4)3 solid electrolyte/electrode interface by polymer interlayer protection," *RSC Adv.*, vol. 10, no. 17, pp. 10038–10045, 2020.

[17] A. L. Davis *et al.*, "Operando X-ray photoelectron spectroscopy of solid electrolyte interphase formation and evolution in Li2S-P2S5 solid-state electrolytes," *Nat. Commun.*, vol. 8, no. 1, pp. 6291–6302, 2020.

[18] M. G. Boebinger, J. A. Lewis, S. E. Sandoval, and M. T. McDowell, "Understanding Transformations in Battery Materials Using in Situ and Operando Experiments: Progress and Outlook," *ACS Energy Lett.*, vol. 5, pp. 335–345, 2020.

[19] L. Yang *et al.*, "Dynamic visualization of the phase transformation path in LiFePO4 during delithiation," *Nanoscale*, vol. 11, no. 38, pp. 17557–17562, 2019.

[20] X. Liu *et al.*, "Distinct charge dynamics in battery electrodes revealed by in situ and operando soft X-ray spectroscopy," *Nat. Commun.*, vol. 4, no. May, pp. 1–8, 2013.

Evidence of Ba-rich surface segregation in $Ba_{1-x}Sr_xTiO_3$ and Ba-rich surfactant in $SrTiO_3$/ $Ba_{1-x}Sr_xTiO_3$ stacks grown by combinatorial pulsed laser deposition

Santiago Agudelo-Estrada[1*], *Nick* Barrett[1], *Christophe* Lubin[1], *Jérôme* Wolfman[2], *Beatrice* Negulescu[2], *Pascal* Andreazza[3], and *Antoine* Ruyter[4]

[1]SPEC, CEA, UMR 3680 CNRS, Université Paris-Saclay, F-91191 Gif-sur-Yvette, France
[2]GREMAN, UMR 7347 CNRS, Univ. de Tours, Parc de Grandmont, F-37200 Tours, France
[3]ICMN, UMR 7374 CNRS, Univ. d'Orléans, 1b rue de la Férollerie, F-45071 Orléans, France
[4]CRISMAT, UMR6508 CNRS, ENSICAEN, 6 Bd du Maréchal Juin, F-14050 Caen, France

Abstract. The interface of a $La_{0.7}Sr_{0.3}MnO_3/SrTiO_3$ bilayer was modulated by introducing 3 unit cells of $Ba_{1-x}Sr_xTiO_3$ using Combinatorial Pulsed Laser Deposition. A wide range of chemical compositions was studied within the same sample, with BSTx stoichiometry variable from 0.5 to 1 along Y-axis, while the SrTiO3 overlayer thickness was modified along the X direction [Fig. 1(a)]. We performed high-resolution, laboratory-based angle-resolved XPS studies of the BSTx film surface providing information on the thickness and composition of the surface and sub-surface layers. Based on the attenuation of the La 3d core-level photoemission signal from the $La_{0.7}Sr_{0.3}MnO_3$ bottom layer, the BST layer is 1.2 nm thick. XPS Ba $3d_{5/2}$ core-level spectra were acquired at positions corresponding to different nominal Ba/Sr stoichiometry. In all measurements, the Ba $3d_{5/2}$ core-level spectra can be represented by two main components, i.e. one component at higher binding energy (BE = 780.54 eV) corresponding to surface contribution and the other one at lower binding energy (BE = 778.92 eV) corresponding to sub-surface contribution (Figs. 2 and 3). Going from normal to 60° emission angle and using a 3-unit cell thick film model, the surface to sub-surface intensity ratio clearly evolves providing evidence of a Ba-rich surfactant. The surfactant effect is more significant for lower nominal Ba stoichiometry.

1 Introduction

The deployment of 5G technology has raised issues of energy consumption, reception quality, and call failure rate. All three can be minimized by continuously adjusting the antenna impedance. A voltage controllable impedance matching circuit with a highly tunable, ferroelectric (FE) capacitance is required. Voltage-tunable capacitors based on the perovskite ferroelectric solid solution $Ba_{1-x}Sr_xTiO_3$ (BSTx) have excellent tunability/losses compromise associated with an easily adjustable Curie temperature (via the Ba/Sr ratio), leading to a far better quality factor (i.e. low dielectric losses at GHz frequencies) than other competing technologies [1, 2]. Working at higher frequencies or lower voltage requires the reduction of the film thickness. In thin films (~50 nm), the formation of a non-FE, chemically distinct interface layer with altered dielectric properties and the increased leakage current can be detrimental to device performance [3]. The idea is to mitigate these limitations by the introduction of an interface control layer that allows manipulating the chemical bonding, promotes atomic rearrangement and possibly engineer better dielectric response. This requires detailed studies of the chemistry and electronic structure of the dielectric/BSTx interface.

2 Experimental

We have used combinatorial pulsed laser deposition to create a library of metal/semiconductor interface layers with continuous in-plane chemical modulation and thickness [4, 5]. We have grown a $La_{0.7}Sr_{0.3}MnO_3/Ba_{1-x}Sr_xTiO_3/SrTiO_3$ sample on a [001] oriented TiO_2 terminated $SrTiO_3$ (STO) substrate, shown schematically in figure 1, with the 3-unit-cell (approximately 1.2 nm) thick BSTx composition modulated along the sample ($0.5 \leq x \leq 1$). The $SrTiO_3$ overlayer thickness varies from 1.1 nm to 15.4 nm. The deposition conditions were: 650°C substrate temperature, 200 mTorr O_2 pressure and 1.6 J/cm² laser fluence. These deposition parameters promote 2D epitaxial growth of the layers, verified through intensity oscillation of Reflection High Energy Electron Diffraction spots before the sample fabrication. The BSTx interfacial layer was deposited from $Ba_{0.5}Sr_{0.5}TiO_3$ and

* Corresponding author: santiago.agudeloestrada@cea.fr

 https://doi.org/10.1051/epjconf/202227301008

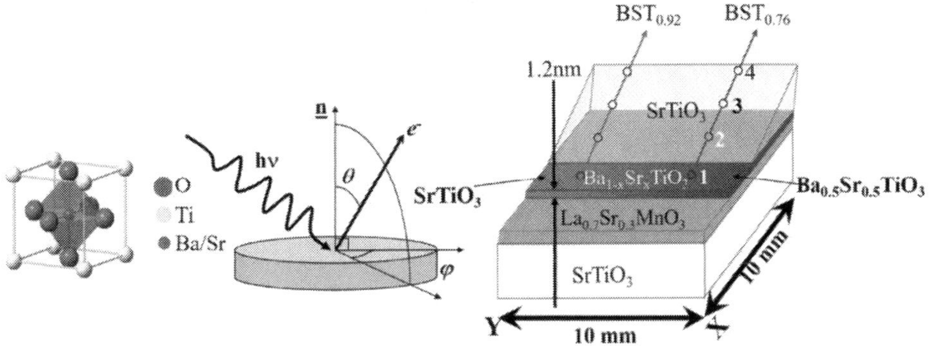

Fig. 1. Schematic of the ferroelectric STO/LSMO (22 nm)/BSTx (1.2 nm)/STO (1.1 nm – 15.4 nm) sample grown by combinatorial pulsed laser deposition and investigated with angle-resolved X-ray photoemission spectroscopy.

$SrTiO_3$ targets. In order to reduce carbon contamination on the surface, the sample was cleaned by short (5 minutes) exposure to ozone just before introduction into the XPS system, followed by annealing in vacuum at 300 °C for 15 min.

The XPS measurements were conducted using a monochromatic Al Kα source (1486.7 eV) and the spectra were recorded with an Argus-128 hemispherical analyzer (ScientaOmicron). The pass energy of 20 eV gave an overall resolution of 0.3 eV. The emission angle relative to the sample surface was varied in order to modulate the probing depth. The deconvolution of the core level spectra was performed using the CasaXPS software [6], and the backgrounds used were of the Shirley type. The peaks were modeled using a Voigt function, a convolution of Lorentzian and Gaussian functions.

3 Results

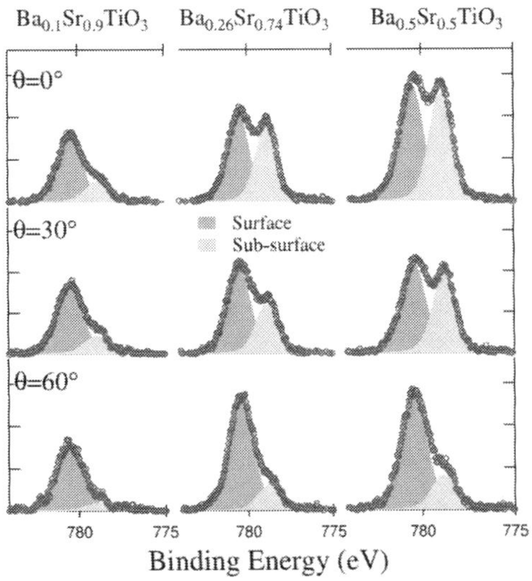

Fig. 2. Ba $3d_{5/2}$ core-level spectra at normal emission (top), 30° (middle), and 60° (bottom) emission angles were acquired for three different BSTx surface positions corresponding to different nominal stoichiometry, i.e. $BST_{0.9}$ (left), $BST_{0.74}$ (center), and $BST_{0.5}$ (right).

We performed high-resolution, laboratory-based XPS studies of the single crystal BSTx film and monocrystalline STO overlayer as a function of the emission angle providing information on the thickness and composition of the surface and sub-surface layers. La 3d and Ba $3d_{5/2}$ core level spectra were recorded. The expected 1.2 nm thickness of the BSTx layer was confirmed by the attenuation of the La photoemission signal from the LSMO bottom layer compared to the signal from the bare LSMO (not shown here).

Figure 2 compares Ba $3d_{5/2}$ core-level spectra acquired at three emission angles from the 1.2 nm thick BSTx layer for three different compositions, i.e. $BST_{0.9}$, $BST_{0.74}$, and $BST_{0.5}$, as a function of the emission angle. Analysis of these spectra reveals two main components: the peak at higher binding energy (BE = 780.54 eV) is characteristic of the surface environment while the peak observed at lower binding energy (BE = 778.92 eV) is characteristic of the sub-surface environment. The relative intensity of the two peaks depends on the probing depth of the analysis (i.e. emission angle) and the nominal Ba/Sr stoichiometry. In fact, the sub-surface Ba intensity becomes lower with respect to that from the surface Ba phase with decreasing probing depth (increasing emission angle) and is, therefore, proportional to the Ba enrichment with respect to the absolute stoichiometry of the underlying BSTx film. By changing the emission angle to 30° and 60°, increasing the sensitivity to the surface with respect to the sub-surface BSTx layer, we observe a relative Ba enrichment of the BSTx surface as compared to the sub-surface.

Table 1. Barium concentration depth profile.

Nominal Stoichiometry	Surface Stoichiometry	Sub-surfaces Stoichiometry
$Ba_{0.5}Sr_{0.5}TiO_3$	0.64	0.43
$Ba_{0.26}Sr_{0.74}TiO_3$	0.36	0.21
$Ba_{0.1}Sr_{0.9}TiO_3$	0.2	0.05

The quantitative analysis of the Ba $3d_{5/2}$ of the BSTx layer points to the segregation of Ba at the surface. Using a 3-unit-cell thick film model (1.2 nm), the ratio of surface intensity to sub-surface [Eq. (1) and (2)] and assuming that the global mean stoichiometry of the 3 layers is the

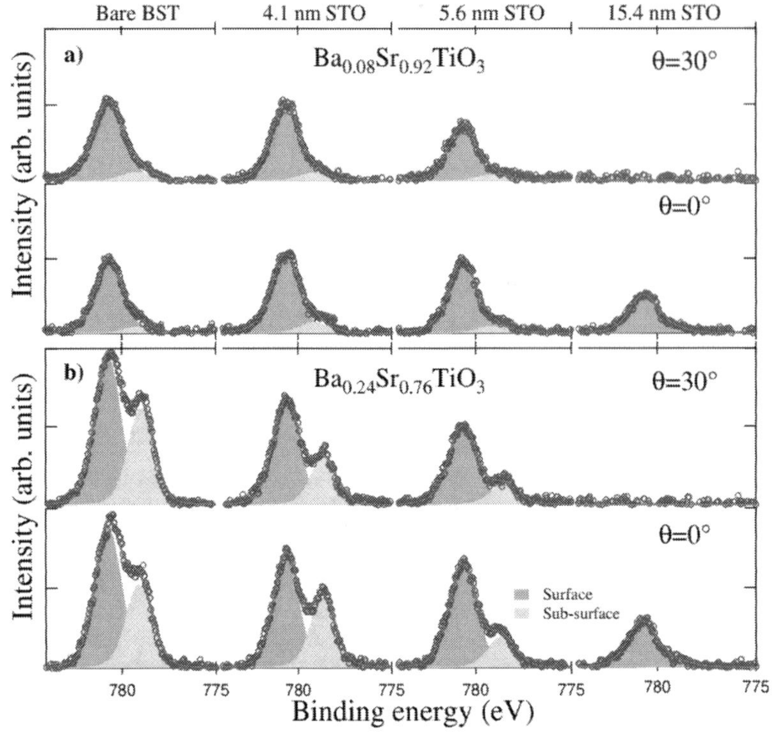

Fig. 3. Ba $3d_{5/2}$ core-level spectra at normal emission (bottom), and 30° (top) emission angles were acquired at four different STO overlayer thicknesses (increases from left to right) at fixed nominal stoichiometry, i.e. (a) $BST_{0.92}$, and (b) $BST_{0.76}$

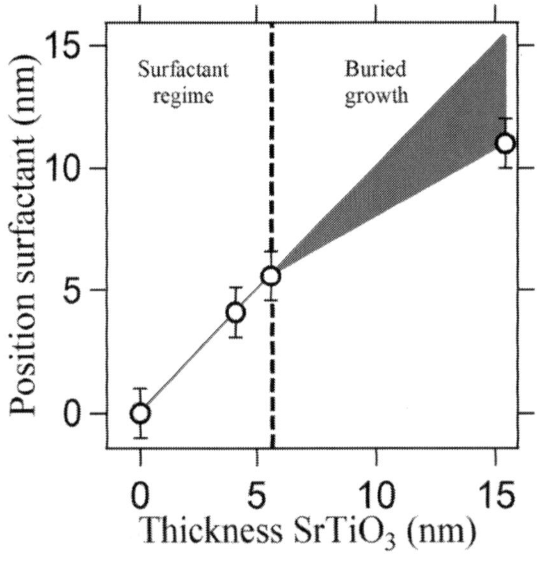

Fig.4. The surfactant position as a function of STO thickness. The suggested dielectric/ferroelectric growth model interface presents a surfactant regime (left) and a buried growth (right).

nominal one; we calculate the barium concentration depth profile [Table (1)]

$$\frac{I_{Surface}}{I_{Sub-surface}} = \frac{Surface\ conc.}{Sub-surface\ conc.(k+k^2)} \quad (1)$$

$$k = exp\left(-\frac{0.4\ nm}{\lambda_{IMFP}\cos\theta}\right) \quad (2)$$

where λ_{IMFP} is the inelastic mean free path of Ba 3d electrons in BSTx And k is the single layer attenuation factor [7]. Its value of 1.5 nm was calculated using the

Tanuma, Powell, and Penn (TPP-2M) algorithm, implemented in the QUASES-IMFP-TPP2M software.

To evaluate the Ba-rich surfactant effect, we performed XPS as a function of the STO overlayer thickness for two different emission angles. Ba $3d_{5/2}$ core-level spectra were recorded. Figs. 3(a) and 3(b) compares the normalized Ba $3d_{5/2}$ core-level spectra from different positions corresponding to four distinct STO thicknesses (0 nm, 4.1 nm, 5.6 nm, and 15.4 nm) at two nominal Ba/Sr stoichiometries, i.e. $BST_{0.92}$ and $BST_{0.76}$.

The most striking feature is the persistence of significant Ba 3d signal up to at least 5.6 nm STO and, in the case of normal emission, up to the thickest STO layer; this is the signature of Ba segregation to the surface of STO, i.e. the formation of a chemically distinct surfactant. Given the estimated inelastic mean free path of 1.5 nm, in the absence of a surfactant effect, we would expect near complete extinction of the Ba 3d emission for 4.1 nm STO overlayer. Instead, the intensity remains comparable to that observed for the bare BST, i.e. there is both Ba segregation at the STO surface and Ba enrichment with respect to the nominal stoichiometry of the underlying BST film.

Analysis of these spectra reveals that the surfactant effect is present up to 5.6 nm. Even for 15.4 nm STO overlayer there is a weak Ba 3d signal at normal emission for both nominal BST stoichiometries. Recalling that the 3λ probing depth is ~ 4.5 nm, we can deduce that the Ba-rich surfactant only starts to be covered by the STO for STO thicknesses $\geq$ 10 nm. This is important information, as thin dielectric layers are often used to modify the electrical boundary conditions, to engineer the depolarizing field or to create diffusion barriers in capacitor or gate stacks.

In Figure 4, we plot the surfactant layer position above the BST film as a function of the STO thickness, assuming that the surfactant layer truly floats on the STO and does not dissipate or dilute into the STO during growth. The surfactant regime ranges from 0 nm to at least 5.6 nm, possibly more. The surfactant layer starts to be buried at higher STO coverages and is ~ (11 $\pm$ 1) nm above the BST film at the end of the STO growth, i.e. buried under 4.5 nm of STO. The driving force behind the Ba surfactant effect may be the epitaxial strain exerted by the STO substrate through LSMO and BSTx layers. Ba effective ionic radii $\left(Ba_{XII}^{2+} = 1.61\text{Å}\right)$ is much bigger than Sr $\left(Sr_{XII}^{2+} = 1.44\text{Å}\right)$ [8] leading to bigger lattice parameters for BSTx compared to STO. Surface Ba with dangling bounds helps reducing the strain, as does Ba atoms progressive dilution into STO matrix.

4 Conclusion

Angle-resolved XPS measurements were performed to measure the surface and sub-surface stoichiometry in a stack suitable for integration into a BSTx based capacitor. We provide evidence of the Barium-rich surfactant effect in BSTx thin films deposited by CPLD. The Ba segregation is stronger for lower nominal Ba/Sr stoichiometry.

A 3-unit-cell thick film model and the ratio of surface intensity to sub-surface obtained by XPS allowed us to calculate the Barium concentration depth profile. The enrichment of Ba content at the surface as compared to the sub-surface may be associated with strain release in the BST epitaxially grown following the smaller lattice parameter of the STO substrate. In addition to the changes in local chemistry, the induced strain gradients may also alter the electrical properties of the BST film, potentially reinforcing local polarization and ferroelectric properties via flexoelectric effect [9].

This project has received funding from the Agence Nationale de la Recherche under Grant No. ANR-20-CE24-0008-BePolar.

References

1. P. Bao, T. J. Jackson, X. Wang, M. J. Lancaster, J. Phys. D: Appl. Phys. **41** 063001 (2008)
2. M. P. J. Tiggelman, K. Reimann, F. Van Rijs, J. Schmitz, R. J. E. Hueting, IEEE Trans. Electron Devices. **56**, 2128-2136 (2009)
3. M. Stengel, D. Vanderbilt, N. Spaldin, Nat. Mater. **8**, 392–397 (2009)
4. C. J. M. Daumont, Q. Simon, E. Le Mouellic, S. Payan, P. Gardes, P. Poveda, B. Negulescu, M. Maglione, J. Wolfman, J. Appl. Phys. **119**, 094107 (2016)
5. J. Wolfman, B. Negulescu, A. Ruyter, N. Niang, N Jaber, Practical Applications of Laser Ablation (2020)
6. N. Fairley, see http://www.casaxps.com/ for complete documentation of fit procedure (2020)
7. L. Q. Zhu, N. Barrett, P. Jégou, F. Martin, C. Leroux, E. Martinez, H. Grampeix, O. Renault, A. Chabli, J. Appl. Phys. **105**, 024102 (2009)
8. R. D. Shannon, Acta Cryst. **A32**, 751-767 (1976)
9. P. Zubko, G. Catalan, A. K. Tagantsev, Annu. Rev. Mater. Res . **43**, 387-421 (2013)

Coupling a magnetic bottle multi-electron spectrometer with a liquid micro-jet device: a comprehensive study of solvated sodium benzoate at the O 1*s* threshold

Marine Fournier[1,2], *Lucie* Huart[2,3,4], *Rémi* Dupuy[1,5], *Régis* Vacheresse[1], *Maximilian* Reinhardt[6], *Denis* Cubaynes[7], *Denis* Céolin[2], *Marie-Anne* Hervé du Penhoat[4], *Jean-Philippe* Renault[3], *Jean-Michel* Guigner[4], *Ajit* Kumar[2], *Bastien* Lutet-Toti[1], *John* Bozek[2], *Iyas* Ismail[1], *Loïc* Journel[1], *Pascal* Lablanquie[1], *Francis* Penent[1], *Christophe* Nicolas[2] and *Jérôme* Palaudoux[1*]

[1] Sorbonne Université, CNRS, Laboratoire de Chimie Physique – Matière et Rayonnement, LCP-MR, F-75005 Paris Cedex 05, France
[2] Synchrotron Soleil, 91192 Saint Aubin, France
[3] Université Paris-Saclay, CEA, CNRS, NIMBE, CEA Saclay, 91191 Gif-sur-Yvette, France
[4] Institut de Minéralogie, de Physique des Matériaux et de Cosmochimie, Sorbonne Université, UMR CNRS 7590, MHNH, 75252 Paris, France
[5] Fritz-Haber-Institut der Max-Planck-Gesellschaft, Faradayweg 4-6, 14195 Berlin, Germany
[6] Nano and Molecular Systems Research Unit, University of Oulu, PO Box 3000, FI-90014, Finland
[7] ISMO, CNRS UMR 8214, Université Paris Sud, bâtiment 350, F-91405, Orsay, France

Abstract. We have developed a magnetic bottle time-of-flight electron-electron coincidence spectrometer to perform measurements on solvated molecules in a liquid micro-jet. We present here the first results obtained after ionization of the oxygen 1*s* inner-shell of sodium benzoate molecules and show the possibilities to filter out the electron signal arising from the liquid phase from the signal of water molecules in the gas phase. Both photoelectrons and Auger electrons spectra (unfiltered and filtered) are presented.

1 Introduction

The interaction of high-energy ionizing radiation with matter is still a very interesting and challenging question in many fields. Radiation damage is a complex multi-scale problem, both in spatial and temporal domains. The interactions begin at the atomic level (atto- to femtosecond time scale), then affect the molecular level (femto- to picosecond time scale), and finally can affect cell behavior over hours or days. The early stages of energy deposition processes are crucial for radiobiological damage, as they initiate electron emission processes [1] whereas X-ray photons can cause complex damage to biological samples [2]. In addition to their possible chemical selectivity, they can eject core (photo)electrons from the molecule. This process can be followed by reorganization/relaxation of the electronic structure (possible ejection of additional Auger electrons). All these electrons are new projectiles with given energies, which can interact with the medium and potentially cause new damages [3]. The investigation of such processes in liquid phase is important because biological systems are surrounded by water molecules, most of the time. Experiments in liquid phase are not readily compatible with vacuum due to the high vapor pressure of water which complicates the application of electron spectroscopy. In addition, the low penetration of soft X-ray photons (few micrometers) and the very short propagation lengths of sub keV electrons in condensed matter (few nanometers) make such studies even more challenging. If in the '70s, electron spectroscopy studies were performed on liquids with a low vapor pressure [4], it is only at the end of the 20th century that Faubel *et al.* studied liquid interfaces using a liquid micro-jet [5]. Since then, many developments occured and liquid micro-jet electron spectroscopy experiments are now common tools [6–9]. Photoelectron spectroscopy provides information about the energies of the emitted electrons, or on the energy dissipation in the solvent (via Coulombic de-excitation or electron transfer phenomena), while photoelectrons / Auger electrons coincidences will provide information on the different relaxation pathways of biomolecules in an aqueous medium and the sharing of energy between the photoelectron and the Auger electrons. This last information may be crucial for increasing the accuracy of the different simulation codes treating the tracks in liquid water like Geant4 [10]. There are several experimental setups where a magnetic bottle – time of flight (MB-TOF) was coupled with a liquid jet (for instance [11,12]), mainly using low repetition rate lasers as ionization sources which could not allow coincidence techniques. Recently, Hergenhahn and co-workers were the first to study electron transfer mediated decay in a lithium chloride solution via electrons coincidences with a MB-TOF using the temporal structure of synchrotron emission [13]. In this work, coincidence measurements were performed by coupling a MB-TOF electron

* Corresponding author: jerome.palaudoux@sorbonne-universite.fr

spectrometer with a liquid micro-jet under vacuum together with soft X-ray synchrotron radiation. Filtering of the data will allow identification of the different Auger structures and a better understanding of the effects of the solvent during de-excitation after inner-shell ionization. The first results obtained above the oxygen K-edge will be presented, allowing the liquid phase signal to be separated from the signal of the surrounding gaseous phase via energy filtering of the photoelectrons. Sodium benzoate molecules were used as target since they have already been studied [14], they are harmless and can be used at relative high concentration (1M and more) which is necessary to get a sufficient photoelectron signal.

2 Experimental description

In this project, we used the experimental setup developed by the LCP-MR, which is an evolution of the earlier MB-TOF experimental setup HERMES (High Energy Resolution Multi Electron Spectrometer) developed to study multiple photoionization processes in the gas phase [15–17]. The spectrometer chamber accomodates a titanium tube (90 mm diameter, 1.2 m-length) on which a solenoid is wound (kapton® insulated copper wire) that generates a homogeneous magnetic field (~0.1 mT). The titanium tube is electrically insulated and can be polarized to ac/decelerate the electrons by applying a suitable electric potential. This parameter may be necessary to improve the energy resolution of the spectrometer. A double-layer mu-metal shielding is placed around the drift tube and the interaction chamber. It avoids Earth and external magnetic perturbations that could deviate the electron trajectories. Finally, two turbo-molecular pumps (600 L/s) are added. One is installed at the entrance of the spectrometer, and the other at the end, near the electron detector. This detector uses micro-channel plates (MCP) coupled with delay lines systems for position encoding (Roentdek®). The overall detection efficiency for one electron is estimated to be ~70% [18] which is directly related to the efficiency of the MCP (active surface). The coupling of the MB-TOF with the liquid jet requires many adjustments because of the permanent evaporation of the sample. Namely, the det ector system must be kept under a high vacuum (< 10^{-5} mbar), while the pressure in the micro-jet chamber can reach 10^{-3}-10^{-4} mbar. This vapor phase can be disruptive for coincidence measurements as it significantly increases the density of events generated in the ionization zone. Moreover, given the low penetration of soft X-rays, only the illuminated part of the liquid micro-jet is probed. The magnet's design was adapted to come as close as possible to the micro-jet without losing magnetic strength in the current setup. For this purpose, a soft iron cone (5 mm base-diameter and 20 mm long) is fixed (by the magnetic force) on top of the permanent magnet (NdFeB truncated cone of 24 mm base-diameter) to concentrate the magnetic field lines. It results in a field of ~0.8 T at the tip of the soft iron cone. This magnet assembly is installed on a XYZ motion controlled by three nano-motors for precise alignment under vacuum. The magnet assembly can be completely removed from the liquid jet holder for the maintenance outside vacuum. Fig. 1 presents a sketch of the setup used in our experiment which is identical to the one described in reference [19]. On the right side of Fig. 1, one can recognize the liquid jet assembly (nozzle and catcher). The conical magnet is placed at a distance of about 1 mm from the liquid jet.

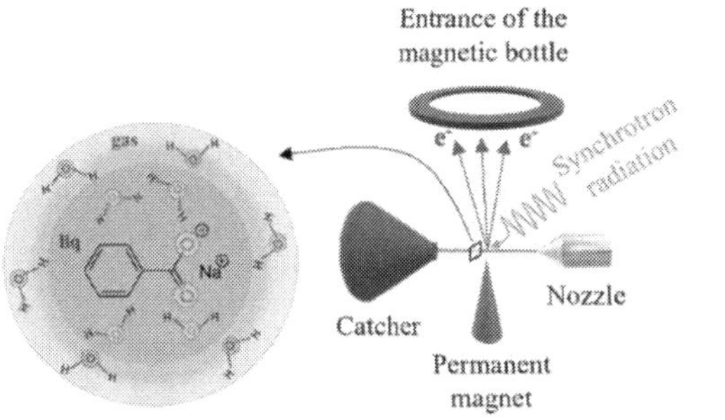

Fig. 1. Scheme of the liquid micro-jet devise coupled with the magnetic bottle electron spectrometer together with a cut view of the liquid and a schematic representation of all types of oxygen atoms present during the measurement.

This assembly is placed in a box that is inserted into the experimental chamber (to maintain the vacuum at a pressure allowing the MCPs to work). The entrance of the magnetic bottle electron spectrometer is located 4 cm above the micro-jet axis and was specially designed with a small 5 mm diameter entrance that allows differential pumping. A manual valve was added to isolate the spectrometer from the experimental chamber in order to decrease the overall venting time when needed.

3 Results at O 1*s* threshold in water (gas and liquid) and discussion

The first measurements were made above the O 1*s* threshold (hν = 580 eV). The electrons are decelerated by 20 V. This allows us to separate the photoelectrons from the huge low-energy electrons signal between 0 and 20 eV (not shown here), to recover resolution and reduce the size of the data files (with such deceleration we are not measuring electrons below 20 eV). hν + M + retardation (-20 V) → M^{+*} (O $1s^{-1}$ gas) + e^-_{ph} (~20 eV). The electrons energies are then reset to their correct values (to ease comparison with literature). The M^{+*} excited states will then decay by Auger relaxation (in the case we consider). Thanks to the magnetic bottle time-of-flight electron spectrometer, we can collect all the electrons emitted during an ionization event and detect in coincidence photoelectron and Auger electron. The two-dimensional coincidence map, presenting the coincidences between photoelectrons (y-axis) and Auger electrons (x-axis) is shown in Fig 2. The number of counts (z-axis) is given in the right part of the figure. We observe here more intense zones corresponding to the true coincidences between two electrons. The total electron spectrum is given on the left side of Fig. 2. The main peak, at 40 eV kinetic energy (blue zone) corresponds to the gaseous water O 1*s* photoelectron whereas the peak at 42 eV kinetic energy (orange zone) corresponds to the liquid water O 1*s* photoelectron. The 2 eV shift between these two photoelectrons' energies is similar to that found in the work of B. Winter [20]. Finally, the faint band at 44 eV (in green) corresponds to the O 1*s* photoelectrons of the benzoate molecule's carboxyl group. The attribution of this peak is consistent with the 1.6 eV shift between the liquid water O 1*s* and the benzoate O 1*s* found in the literature [8]. The low intensity of this band is easy to understand: one molar concentration means that one sodium benzoate molecule is surrounded by 55 water molecules. Sodium benzoate is an amphiphilic molecule, so we do expect that there should be a higher benzoate-to-water ratio at the surface. However, previous studies on sodium benzoate using conventional XPS [14,21] observed that the signal of sodium benzoate remains significantly weaker than that of benzoic acid, its protonated counterpart. It was thus concluded that the surface propensity of sodium benzoate is not so high, thus we may expect an increase of a factor of a few compared to the bulk ratio but not more than an order of magnitude. A precise quantification has not been performed. The observed ratio between the liquid and the gaseous peaks can be explained since the electrons with kinetic energies of a few tens of eV can escape from only a few nm. Other considerations, for instance light polarization should be taken in account.

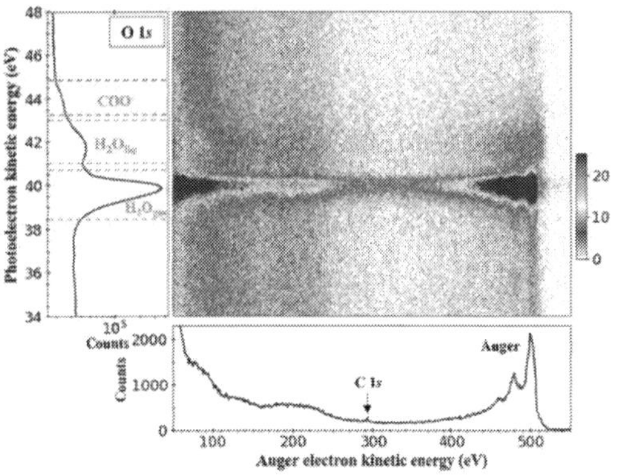

Fig. 2. Sodium benzoate two-dimensional coincidence map « Auger electron kinetic energy / photoelectron kinetic energy » (middle part). The photoelectron spectrum is shown on the left, with a step of 0.1 eV. The total (unfiltered) Auger electron spectrum (lower part, in black) is shown with a step of 1 eV.

The total spectrum (in kinetic energy) obtained by projecting the 2D map data onto the x-axis is given in the lower part of Fig. 2. In this spectrum we can recognize the Auger bands of O 1*s* between 450 eV and 510 eV. These bands are well known but their shape is different from similar spectra in the literature. This point will be discussed later in the manuscript. In this spectrum, one can also point out a peak around 280 eV which is the C 1*s* photoelectron signature. Associated with this peak we distinguish a broad structure between 150 eV and 270 eV which can be associated (energetically) with the C 1*s* Auger electrons. Such structures can also be associated with secondary electrons generated after oxygen atoms inner-shell ionization, from collisions and from other decay processes (as cascades). Nevertheless, this total electron energy spectrum does not allow to distinguish the Auger relaxations consecutive to the emission of one O 1*s* photoelectron (from gaseous or liquid water or a molecule). Our coincidence measurements allow to select an energy range within the data corresponding to the photoelectron of interest. Projecting this filtered data along the x-axis gives the filtered Auger spectra in Fig. 3.

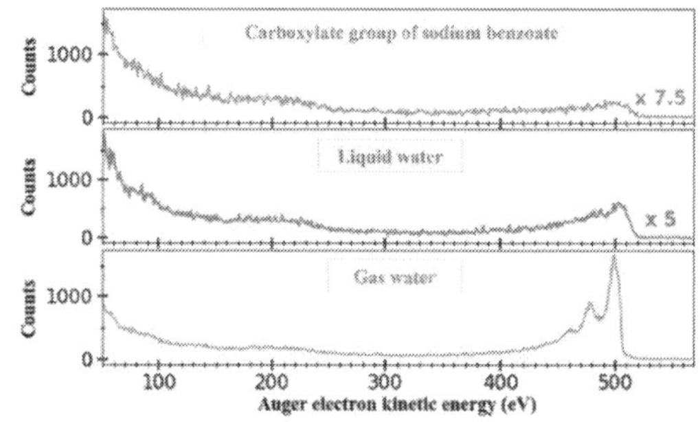

Fig. 3. Photoelectron-filtered spectra with a step of 1 eV, showcasing Auger features. The colors of the spectra refer to the nomenclature of the different oxygens species given in Fig 2. Scaling factors were applied to the liquid and molecular spectra for comparison.

Fig. 3 shows the filtered Auger spectra (in kinetic energy) after ionization of the O 1*s* atom in gaseous water (bottom part of Fig. 3, in blue), liquid water (middle part of Fig. 3, in orange), and in the benzoate molecules (upper part, in green). The colors of the spectra correspond to those chosen for the photoelectrons in Fig. 2. In the gaseous water spectrum (in blue), between 450 eV and 510 eV, one can clearly distinguish the well-known Auger bands of O 1*s* comparable to those in the literature ([22,23] for example). The Auger spectrum from water in the liquid phase (orange) is less clear. Nevertheless, similarities are found such as diminution of the peaks resolution and a 10 eV shift with the main O 1*s* gaseous Auger band as seen by Öhrwall in water clusters [24] suggesting that such media begin to behave like a liquid. Our spectrum remains identical to [8,20] which were obtained after subtraction of the gas phase contribution. However, our coincidence method allows a direct and selective measurement, preventing having data analysis artifact when removing the gas-phase contribution. In the case of benzoate, there is no comparison available in the literature (gas phase, liquid or solid). In the benzoate spectrum, a smaller band is visible in the same energy range as the O 1*s* Auger of liquid water suggesting a similar behavior. On the other hand, a factor of 3 can be seen between the spectrum of gaseous water and the spectra of liquid water and molecule. Similar studies can also be performed at other edges (inner-shells) like C 1*s* for instance. The main purpose for it is also that only the solvent (water in our case) can evaporate from the liquid jet. The observation of the C 1*s* electron signal is another clear sign of the liquid medium, as described in [19].

4 Conclusion and perspectives

We report here the first results from a new magnetic bottle time of flight spectrometer dedicated to electron spectroscopy of solvated molecules in the liquid micro-jet environment using the electron-electron coincidence technique. In the O 1*s* inner-shell ionization case we have shown that it is possible to distinguish events from the gaseous water phase and the liquid phase (water or benzoate molecules) in the photoelectron spectrum. Strong theoretical support is needed to disentangle the different Auger spectra. After an inner-shell initial vacancy, different de-excitation pathways can occur such as Inter Coulombic Decay, Electron Transfer Mediated Decay, and classical Auger decays (normal + resonant). Cascade processes can also occur within the « normal » Auger decay, where several secondary electrons are emitted (once a more ionized final state threshold, M^{3+}, M^{4+}, … is energetically open). The interactions between the molecules of interest and the molecules of the solvent must be studied in detail. Several molecules of biological interest at different shells (O 1*s*, C 1*s*, N 1*s*, …) could be considered with our coincidence method to obtain cross sections that will be useful as benchmarks for future theoretical studies.

Acknowledgments

Experiments were carried out on PLEIADES beamline with the approval of Synchrotron SOLEIL (proposals 20190766 and 20191641). We thank the SOLEIL biology and chemistry laboratories for their helpful support, as well as E. Robert from PLEIADES beamline for his technical support. This work was supported by the ANR project HighEneCh (ANR-17-CE30-0017), the NanoTheRad strategic Research Initiatives from Paris Saclay University and the IraSolubiox project from Labex PALM (ANR-10-LABX-0039-PALM). J.P. and R.V. thank H. Ringuenet from LCP-MR mechanical workshop. J.P., F.P. and C.N. acknowledge U. Hergenhahn for fruitful discussions and advises.

Author contributions

C.N. and J.P. designed the research. All authors performed research. L.H., M.F., P.L. I.I. and J.P. analysed the data. M.F., L.H., F.P., I.I., C.N., R.D. and J.P. wrote the paper.

References

1. A. Tilikidis and A. Brahme, Acta Oncologica **33**, 457 (1994).
2. R. Cox, J. Thacker, D. T. Goodhead, W. K. Masson, and R. E. Wilkinson, International Journal of Radiation Biology and Related Studies in Physics, Chemistry and Medicine **31**, 561 (1977).
3. E. Alizadeh, T. M. Orlando, and L. Sanche, Annu. Rev. Phys. Chem. **66**, 379 (2015).
4. H. Siegbahn and K. Siegbahn, Journal of Electron Spectroscopy and Related Phenomena **2**, 319 (1973).
5. B. Winter and M. Faubel, Chem. Rev. **106**, 1176 (2006).
6. R. Dupuy, C. Richter, B. Winter, G. Meijer, R. Schlögl, and H. Bluhm, J. Chem. Phys. **154**, 060901 (2021).
7. M. Ammann, L. Artiglia, and T. Bartels-Rausch, in *Physical Chemistry of Gas-Liquid Interfaces* (Elsevier, 2018), pp. 135–166.
8. M. A. Brown, M. Faubel, and B. Winter, Annu. Rep. Prog. Chem., Sect. C: Phys. Chem. **105**, 174 (2009).
9. R. Signorell and B. Winter, Phys. Chem. Chem. Phys. **24**, 13438 (2022).
10. S. Agostinelli *et al.*, Nuclear Instruments and Methods in Physics Research Section A: Accelerators, Spectrometers, Detectors and Associated Equipment **506**, 250 (2003).
11. N. Kurahashi, S. Thürmer, S. Y. Liu, Y. Yamamoto, S. Karashima, A. Bhattacharya, Y. Ogi, T. Horio, and T. Suzuki, Structural Dynamics **8**, 034303 (2021).
12. I. Jordan, A. Jain, T. Gaumnitz, J. Ma, and H. J. Wörner, Review of Scientific Instruments **89**, 053103 (2018).
13. M. N. Pohl, C. Richter, E. Lugovoy, R. Seidel, P. Slavíček, E. F. Aziz, B. Abel, B. Winter, and U. Hergenhahn, The Journal of Physical Chemistry B **121**, 7709 (2017).
14. N. Ottosson, A. O. Romanova, J. Söderström, O. Björneholm, G. Öhrwall, and M. V. Fedorov, J. Phys. Chem. B **116**, 13017 (2012).
15. H. Iwayama, N. Sisourat, P. Lablanquie, F. Penent, J. Palaudoux, L. Andric, J. H. D. Eland, K. Bučar, M. Žitnik, Y. Velkov, Y. Hikosaka, M. Nakano, and E. Shigemasa, The Journal of Chemical Physics **138**, 024306 (2013).
16. F. Penent, J. Palaudoux, P. Lablanquie, L. Andric, R. Feifel, and J. H. D. Eland, Physical Review Letters **95**, (2005).
17. J. Palaudoux *et al.*, Physical Review A **82**, (2010).
18. K. Fehre *et al.*, Review of Scientific Instruments **89**, 045112 (2018).
19. L. Huart *et al.*, Physical Chemistry Chemical Physics (2022).
20. B. Winter, E. F. Aziz, U. Hergenhahn, M. Faubel, and I. V. Hertel, The Journal of Chemical Physics **126**, 124504 (2007).
21. L. Huart, Inner Shell Ionization Effects on Molecules of Biological Interest in an Aqueous Medium, Sorbonne University, 2022.
22. W. E. Moddeman, T. A. Carlson, M. O. Krause, B. P. Pullen, W. E. Bull, and G. K. Schweitzer, The Journal of Chemical Physics **55**, 2317 (1971).
23. L. Inhester, C. F. Burmeister, G. Groenhof, and H. Grubmüller, The Journal of Chemical Physics **136**, 144304 (2012).
24. G. Öhrwall *et al.*, The Journal of Chemical Physics **123**, 054310 (2005).

On the energy resolution of a GaAs-based electron source for spin-resolved inverse photoemission

Abraham Federico Campos[1*], *Thomas* Duden[2], and *Antonio* Tejeda[1]

[1]Laboratoire de Physique des Solides, 91400 Orsay, France.
[2]Constructive Solutions for Science and Technology, 33649 Bielefeld, Germany.

Abstract. The spin resolution in inverse photoemission spectroscopy is achieved by injecting spin-polarized electrons, usually produced by GaAs-based cold cathodes that replace hot-filament electron guns of spin-integrated setups. The overall energy resolution of the system can be enhanced by adjusting either the optical bandpass of the photon detector or the energy distribution of the electron beam. Here we discuss the influence of the photocurrent and the photocathode temperature on the energy broadening of the electron beam through the inverse photoemission spectra of the spin-splitted Shockley surface state of Au(111). First, we find that cooling down the GaAs photocathode to 77 K increases the band gap and reduces the number of allowed vertical transitions, monochromatizing the electron beam with an enhancement of about 30 meV for the energy resolution. Second, we observe a correlation between the generated photocurrent at the electron source, and the space-charge effects at the sample as a reduction of lifetime and spin asymmetry of a polarized bulk state. These observations allow defining a threshold of current density for the optimum acquisition in the measurements of spin-resolved inverse photoemission in Au.

1 Introduction

The exploration of unoccupied states with total spin control is now possible in inverse photoemission (IPES) by fully decoupling the spin polarization vector of the electron beam from its wavevector [1]. The decoupling of the polarization orientation of the electron beam from the wavevector allows exploring the unoccupied bands with the spin orientation more suitable to explore every state. Another critical factor when studying the bands is the energy resolution in order to distinguish states close in energy. It has already been demonstrated that spin-resolved measurements of spin-polarized inverse photoemission (SPIPES) allow discriminating unoccupied spin-dependent states whose energy splitting is below the energy resolution [2]. Any enhancement of the total energy resolution ΔE_{tot} in IPES setups will therefore allow to better discriminate states close in energy, as required in particular for the study of spin-textures in systems with e.g., spin-orbit coupling (SOC), magnetic exchange or Rashba effect. The total energy resolution depends on the thermal distribution of the electron beam and the bandpass energy of the photon detector with usual detection in the vacuum ultraviolet (VUV) regime. In isochromat IPES, where only the kinetic energy of the electron beam is varied, the bandpass detection of the emitted photons consists of a combination of the photoionization threshold of the detection gas (high-pass) and the transmission cutoff of an optical window (low-pass) in Geiger-Müller counters. A modification of the bandpass characteristic through the decrease of the low-pass energy transmission allows for enhancement of the optical resolution of the photon detector system. The surface quality and purity [3], and the temperature of the window [4,5], usually made of alkaline earth fluoride crystals, are the most explored parameters. The temperature dependency of the cutoff wavelength of VUV transmission windows, explained by a photoexcitation model of excitons at thermally distorted lattice regions [5,6], allows to reach up to 165 meV for the overall energy resolution [4]. In more elaborated devices the high-pass is modified by using the oxygen and krypton absorption lines by incorporating additional windows [7]. However, the increase of energy resolution by the aforementioned methods is usually accompanied by a decrease in the quantum yield [8] which is quite disadvantageous due to the intrinsic cross-section of IPES. Thus, an alternative to enhance the energy resolution is by adjusting the beam distribution of the GaAs-based electron source. Only a few studies deal with the photoemission broadening of the polarized electrons, see for instance [9,10]. Varying the temperature of the electron source rather than the temperature of the window of the photon detector is an alternative that does not reduce the counting rate in IPES since it does not filter the energy of the beam electrons. For these reasons, we report here the effects of the temperature of a GaAs-based electron source on the energy resolution in isochromat IPES by analyzing the spin-polarized Shockley surface state of Au(111). From

* Corresponding author: campos-contreras@lps.u-psud.fr

the phase space considerations of the IPE process, it would be desirable to have a high-intensity source with a small beam size and high flux. Yet, even when samples could stand relatively high currents, the electron-electron repulsion becomes important in low-energy and high-brightness beams, affecting not only the energy but also the momentum resolution [11,12]. The selected transfer energy in the electron optics has been chosen to reduce them [1]. Therefore, the effects of the space-charge are observed as a decrease in the lifetime of the surface state while increasing the electron current at the target.

2 Methodology

The Au(111) surface was prepared by Ar ion sputtering (1 keV) and annealing at 800 K under the UHV environment. The cycles were repeated until a clear LEED pattern with sixfold symmetry was obtained (not shown). The SPIPES measurements were subsequently performed.

The spin-resolution is incorporated in the room temperature measurements on freshly-prepared samples by the near-infrared photoemission (~830 nm) from a negative electron affinity GaAs wafer, already detailed for isochromat IPES [1]. The net polarization of the photocathode is reported to be $P = 0.30 \pm 0.03$. The photocathode was cooled down to LN_2 boiling point (77 K) through an in-house cold trap at the 3-axis photocathode manipulator and stabilized in about 1 hr. The temperature is monitored by a contact thermocouple at the photocathode surface. Because thermal fluctuations of the GaAs lattice modify the sweet spot of the NIR optical excitation, impacting the transmission of the electron beam to the sample during the cooling process, the experiment was performed until the electron transmission (~55%), and the photocathode temperature stabilized.

3 Results

The degree of polarization of electronic states due to SOC is more significant in heavy-atom compounds and it can be enhanced by surface adsorbates as occurs in noble-metal surfaces with Rashba splitting [13]. However, if the Rashba parameter is very small, the spin effects may not be observed in IPES due to the constraints on the energy resolution, usually of hundreds of meV. A prototypical system for studying the Rashba splitting is the Shockley surface state of Au, studied below [14-16] and above the Fermi level [1,17], with a spin that is tangent to the concentric Fermi surface. We present in Fig. 1 the SPIPES raw spectra of Au(111) along the ΓM direction at room temperature. The spectra are normalized to the impinging current of ~ 0.6 μA. The spin-polarized Shockley surface state of the L-gap and surface-projected bulk bands are observed, and the

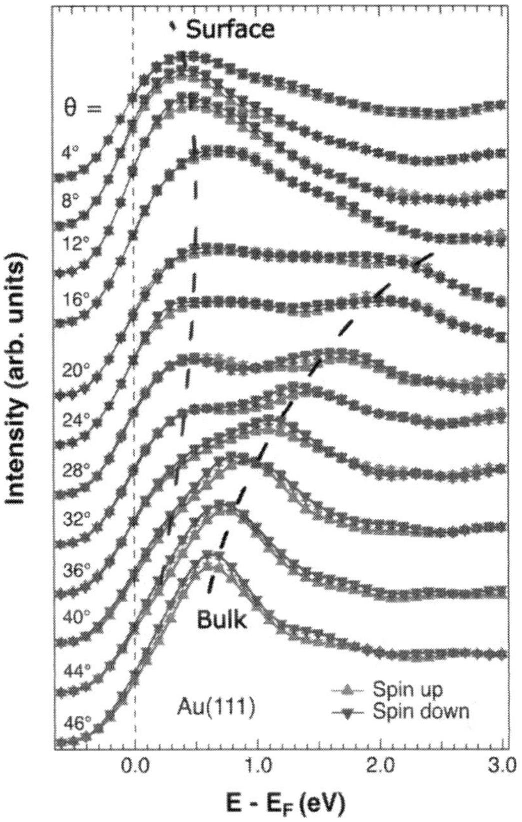

Fig. 2. SPIPES spectra of Au(111) along ΓM at room temperature. Surface and bulk states are present. Lines are added as a guide to the eye.

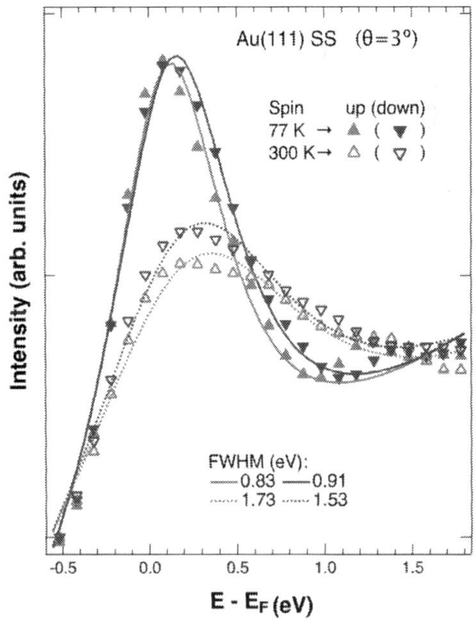

Fig. 1. SPIPES spectra of Shockley surface state of Au(111) at θ = 3° as a function of the temperature of the GaAs photocathode. In the LT, the binding energy of the state is more defined as follows from the FWHM.

details on the surface resonance and spin-polarized bulk states have been discussed elsewhere [1,17]. The lifetime of the Shockley surface state is apparently smaller when approaching normal incidence due to the particular experimental constraints. It is thus desirable to increase the energy resolution to better estimate the spin-split final state binding energies. Therefore, we focus first on the thermal distribution of the electron beam while studying the surface state at an almost normal incidence (θ=3°), as shown in Fig. 2.

Two spectra are contrasted as a function of the photocathode temperature: the 77 K low temperature (LT) and the 300 K high temperature (HT) spectra. Both measurements are normalized to a target current kept at about 0.6 μA for both sets of measurements. The system presents the surface Rashba effect so that the binding energy of the polarized final states are shifted in energy with respect to each other for a given wavevector. It is evident that the IPES spectral linewidth of the Shockley surface state of Au(111) is decreased when the GaAs temperature goes from HT to LT, facilitating the determination of the binding energy of the state. The difference of the full width at half maximum (FWHM) is 0.9 (0.6) eV for the spin up (down) component. By increasing the bandgap energy of GaAs, the optical transitions are then more concentrated around the Γ-point as indicated by the rise of the spectral intensity in the LT surface-emission while maintaining the same inelastic background of the HT at ~1.5 eV. In this scenario, the effective polarization of the GaAs photocathode should also increase due to the monochromatized electron beam. A difference in the binding energy is possibly due to a variation of the photocathode work function at these two temperatures. Experimentally, this variation can be compensated by modifying the acceleration potential of the electron source [1]. Anyway, here we are concerned with the broadness of the SPIPES features. The SPIPES spectral broadening is mainly affected by two parameters: the thermal energy spread of the beam and the space-charge effects of the cathode. If we concentrate at first on thermal effects, the broadening when considering a Maxwellian momentum distribution projected along one transverse direction is given by [11, 18]:

$$\Delta k(T) = \frac{2}{\hbar}\sqrt{2mk_BTln(2)}, \quad (1)$$

where T is the photocathode temperature. Since electrons arriving at the sample are free electrons, their energy broadness is:

$$\Delta E = \frac{\hbar^2}{m}k\Delta k. \quad (2)$$

And combining Eq.1 and Eq.2 we get that

$$\Delta E(k,T) = \frac{2\hbar}{m}k\sqrt{2mk_BTln(2)}. \quad (3)$$

If only the photocathode temperature is varied, the change between energy distributions can be quantified at each wavevector k_i. Assuming that the binding energy of the spin-down state is $E^{\downarrow} \sim 0.2$ eV, we get that $k_{\parallel} \sim 0.06$ Å^{-1} and therefore $\Delta E(300) - \Delta E(77) = 62 - 32$ meV $= 30$ meV. In other words, there is a difference of 30 meV in the thermal distribution of the electron beam when the photocathode is at 77 K.

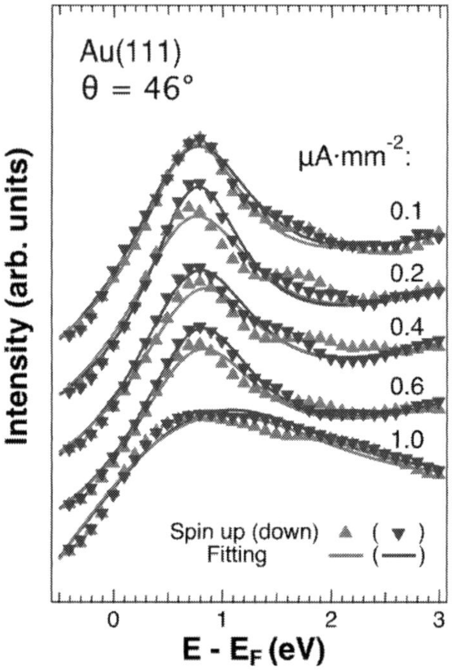

Fig. 3. SPIPES spectra of Au(111) *sp*-bulk state (θ = 46°) as a function of the current density over the sample. The data (triangles) are normalized to maximum intensity and fitting lines are shown. The effects of the current density are observed in: (i) the spin asymmetry and (ii) the spectral broadening of the state.

The above energy resolution of the inverse photoemission setup has been determined by avoiding the space-charge effect. The space-charge at the photocathode surface may broaden the thermal energy distribution of the electron source if the emitted current is too high. This eventuality is usually prevented by decreasing the voltage in the extractor lenses at the electron source. Experimentally, we studied the space-charge effects by varying the impinging current into the *sp*-bulk state (θ = 46°) of the Au(111) surface. This state presents spin asymmetry and pseudo-Rashba splitting as experimentally observed and ascribed to initial-state effects [15]. The SPIPES spectra of the state as a function of current density are presented in Fig. 3 and their corresponding FWHM are presented in Fig. 4. The space-charge affects the electron beam for relatively high current densities over the sample. It can be seen a gradual increment of the FWHM of the state as a function of the current density, up to the point that it is difficult to determine the final state binding energy when the beam current approaches 1.0 μA·mm^{-2}. Thus, it is critical to characterize the current density over the specific sample since this may lead to misinterpretations of the lifetime of the state. This is why we have determined the most favorable current densities for exploring Au(111) with SPIPES and in the range between 0.2 and 0.6 μA·mm^{-2}.

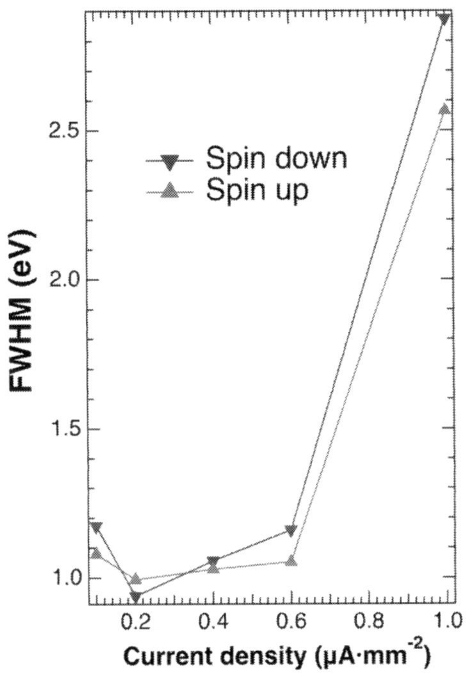

Fig. 4. FWHM of the SPIPES spectra of Au(111) *sp*-bulk state (θ = 46°) as a function of the current density over the sample.

In summary, cooling down the photocathode to 77 K increases the energy resolution attributed to the electron beam broadening by 30 meV while not reducing the electron beam intensity. This study has been performed by previously determining experimental conditions where the space-charge effects are negligible for the bulk state of Au(111).

The authors strongly acknowledge CNRS (Institut de Physique) for funding the spin- and angle-resolved inverse photoemission project and Labex Palm–Grant No. ANR-10-LABX-0039-PALM–for extra funds (SPOUSE and IP-OP projects). We also acknowledge funding by DIMAG project from 2019 FLAG-ERA ANR call.

References

1. A. F. Campos, S. Cabaret, P. Duret, T. Duden, A. Tejeda, Rev. Sci. Instrum., **93**, 093904 (2022).
2. M. Donath, App. Phys. A, **49**, 351-364 (1989).
3. C. Thiede, A. B. Schmidt and M. Donath, Rev. Sci. Instrum., **86**, 085101 (2015).
4. M. Budke, V. Renken, H. Liebl, G. Rangelov, M. Donath, Rev. Sci. Instrum., **78**, 083903 (2007).
5. W. R. Hunter and S. A. Malo, J. Phys. Chem. Sol., **30**, 2739-2745 (1969).
6. M. H. Reilly, J. Phys. Chem. Sol., **31**, 1041-1056 (1970).
7. R. Stiepel, R. Ostendorf, C. Benesch and H. Zacharias, Rev. Sci. Instrum., **76**, 063109 (2005).
8. F. Shedin, G. Thornton and R. I. G. Uhrberg, Rev. Sci. Instrum., **68**, 41 (1997).
9. H. J. Drouhin, C. Hermann, G. Lampel, Phys. Rev. B, **31**, 3859 (1985).
10. H. J. Drouhin, C. Hermann, G. Lampel, Phys. Rev. B, **31**, 3872 (1985).
11. N. G. Stoffel and P. D. Johnson, Nucl. Instrum. Meth. Phys. Res. A, **234**, 230-234 (1985).
12. M. Maniraj and S. R. Barman, Rev. Sci. Instrum., **85**, 033301 (2014).
13. H. Mizushima, H. Koike, K. Kuroda, Y. Ishida, M. Nakayama, K. Mase, T. Kondo, S. Shin and K. Kanai, Phys. Chem. Chem. Phys., **19**, 18646-18651 (2017).
14. S. LaShell, B. McDougall, E. Jensen, Phys. Rev. Lett., **77**, 3419 (1996).
15. M. Hoesch, M. Muntwiler, V. N. Petrov, M. Hengsberger, L. Patthey, M. Shi, M. Falub, T. Greber and J. Osterwalder, Phys. Rev. B, **69**, 241401 (2004).
16. F. Reinert and G. Nicolay, Appl. Phys. A, **78**, 817-821 (2004).
17. S. N. P. Wissing, C. Eibl, A. Zümbulte, A. B. Schmidt, J. Braun, J. Minár, H. Ebert and M. Donath, New J. Phys., **15**, 105001 (2013).
18. W. Sinz, Nucl. Instrum. Meth. Phys. Res., **187**, 259-262 (1981).

https://doi.org/10.1051/epjconf/202227301011

Photoemission in microelectronic research: When technological developments allow to tackle new questions in the lab.

Thierry Conard
Imec/MCACSA, Kapeldreef 75, 3001 Leuven, Belgium

Abstract. Photoemission has always played an important role in the development of new microelectronic devices and processes and has quickly focused on the use of monochromatized Al Kα radiation in most commercial instruments. While, for instance, ARXPS using sample tilting or the use of high-energy photoemission at synchrotron facilities were available, their use remained confidential due to technical constraints such as measurement time, repeatability of the measurements and/or access to the facilities. In this paper, we show through some examples that the implementation of these features in a user-friendly setup in the laboratory considerably enhances the information retrieved from photoemission experiments.

1 Introduction

Since the development of the transistor by Bardeen and Brattain[1] in 1947, the semiconductor technology has been evolving considerably. Since the introduction of integrated circuits, performances have been progressing according to Moore's law[2] which was formulated in 1965. For many years, performance improvements were achieved primarily by geometrical scaling and the technology nodes were named according to the gate length of the transistor. However, facing physical limitations, from the 45 nm node no further scaling of the gate length has occurred[3] and at the 22 nm node Intel introduced the FinFET technology shifting away from planar technology. But even within the planar technology, scaling was only possible through a different choice of materials such as the replacement of SiO_2 as gate oxide by oxides with higher permittivity or by the introduction of copper in interconnects. These technological evolutions also required improved capabilities in material characterisation.

This article focuses on the contribution of photoemission to the technological evolution of transistors and more specifically on the interest of Angle-Resolved X-ray Photoelectron Spectroscopy (ARXPS) and acronym Hard X-ray Photoelectron Spectroscopy, (HAXPES) to these technological developments. Both variants of PES are known for a very long time, going back to the earliest publications on XPS but were not widespread due to practical difficulties. ARXPS for instance required sequential measurements at various sample tilts leading to lengthy measurement times and difficult sample alignment. The introduction of parallel-ARXPS, allowing to get angular photoelectron information over a wide range of angles in a single measurement alleviates this limitation and is implemented by several manufacturers. Similarly, HAXPES required access to a synchrotron facility for the use of high intensity monochromatic sources, which is impractical for process development and more routine analysis. The situation changed radically when high energy monochromatic sources with photon energies ranging from ~2.98 keV to 9.25 keV were brought to the laboratory on standard XPS platforms.

2 Experimental

The photoemission results presented in this article are originating from three different instruments.

For ARXPS, all measurements were performed on a Theta 300 spectrometer from Thermo Fisher, using Al Kα radiation at 1486.6 eV. The exit angular range of the detected photoelectrons was between 20 and 80 degrees, as measured from the normal of the sample and intensities were binned by ~4 degrees.

For XPS, all measurements were performed either on a Physical Electronics Quantes system using Cr Kα radiation at 5414.8 eV or on a ScientaOmicron HAXPES-lab system with Ga Kα radiation at 9250 eV.

3 A first paradigm shift: high-k metal gate

One of the challenges faced by geometrical scaling is that the reduction of the gate oxide thickness led to an increase of leakage current due to tunnelling. This high leakage became unacceptable due to high electric consumption, which is even more critical for mobile applications. This can be improved by using higher permittivity materials which allow a larger physical thickness while keeping a low Equivalent Oxide Thickness (EOT). At first this was achieved by incorporating nitrogen into SiO_2, which allowed keeping the technology of *grown oxide* but the limitation of the increase of permittivity led to the use of *deposited oxides* (Al_2O_3, ZrO_2, HfO_2, ...) through major technological developments.

https://doi.org/10.1051/epjconf/202227301011

3.1 Growth quality of ALD grown layer

When switching from a process of grown oxide to deposited oxide, a fundamental question is the quality of the deposited oxide in terms of uniformity both at the microscopic and macroscopic level. Indeed, very quickly, Atomic Layer Deposition (ALD) became the method of choice due to its good conformality and growth control. However, as a surface driven process, substrate surface preparation is critical and the growth quality in function of surface termination needs to be evaluated.

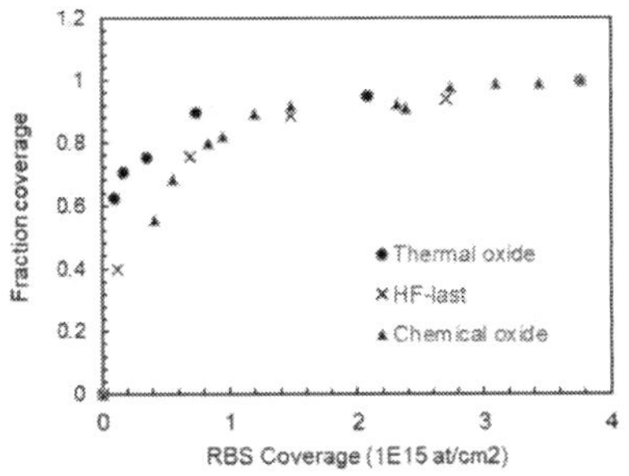

Figure 1 : Evolution of the fractional coverage of Al_2O_3 layer grown on various Si surfaces as a function of deposited material as measured by RBS.

The evaluation of the growth quality in the first deposition cycle needs a quantitative surface analysis technique with sufficient sensitivity. Techniques that have been used/considered include Low Energy Ion Scattering (LEIS), Time-Of-Flight Secondary Ion Mass Spectrometry (ToF-SIMS), XPS and RBS (Rutherford Back Scattering), in decreasing order of surface sensitivity. As is often the case, no single technique matches all criteria. LEIS is extremely surface sensitive and a quantitative technique but has issues with detection limits due to the low counting rate. ToF-SIMS is very sensitive but has issues with quantification due to the matrix effect. XPS, has in all aspects (detection limits, surface sensitivity, quantification accuracy) characteristics similar or in between LEIS and ToF-SIMS and can thus be an adequate technique for the evaluation of the growth quality of ALD layers.

One of the limitations of photoemission is that quantification in complex systems relies on intensity modelling. This has been done for instance in the case of the deposition of Al_2O_3 on Si (100) with various surface preparation[4]. In that case intensity modelling was done assuming that the layers have grown in a nonuniform manner, producing island structures on the silicon dioxide. The parameters used in the modelling are the thickness of the islands and the fractional coverage. Further, it is assumed that the islands are such that shadowing of the signal emitted from the uncovered SiO_2 is not significant. In that case, it shown that the angular dependence of the Al 2p and Si 2p intensities is very sensitive to variation in the combination of the coverage and thickness of the islands. Using this method, a clear distinction of the growth of Al_2O_3 on HF-terminated Si, a thin thermal SiO_2 layer or a chemical surface oxide has been observed (Figure **1**). It should however be mentioned that, due to its better detection limits, ToF-SIMS allows a better discrimination of the growth quality when the surface conditions are very similar, and that the matrix effect can be neglected[5].

3.2 Determining composition profiles of thin layers

When good quality layers can be grown, additional parameters need to be characterised. First of all, layer thickness is critical for devices performances. For single layer on a substrate, in theory, a XPS measurement of the intensities of both the substrate and the overlayer at a single angle are sufficient for determining layer thickness. This gives excellent results and can even be considered superior to traditional ellipsometry measurements for very thin layers, the latter technique cannot discriminate between the contamination layer and the deposited layer. An example of the thickness measurement of a HfO_2 layer deposited on SiO_2 correlated to the amount of deposited material as measured by RBS is presented in Figure **2**.

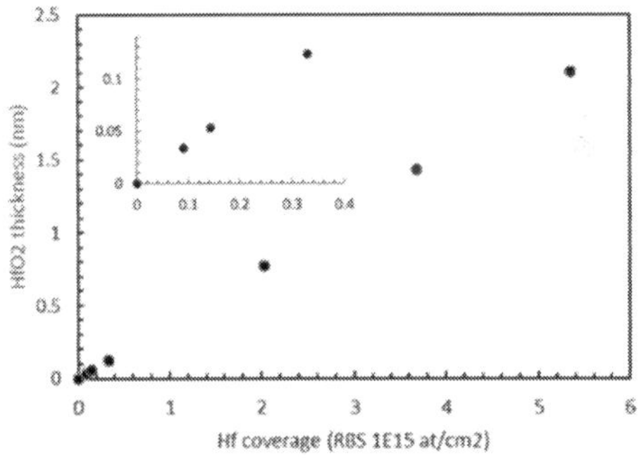

Figure 2 : Thickness calculation based on XPS data of a HfO_2 layer deposited on SiO_2 correlated to the amount of deposited material as measured by RBS

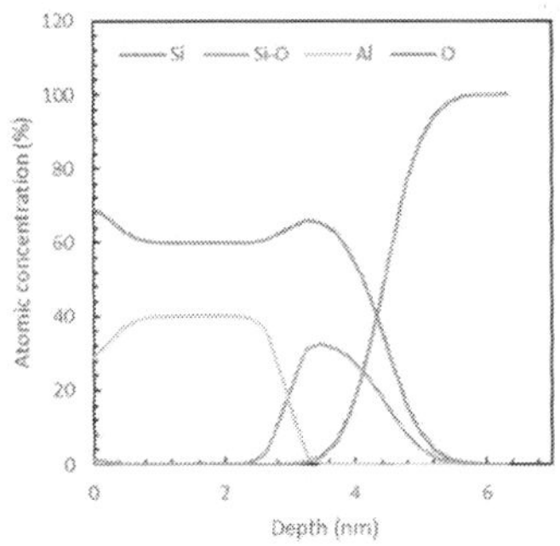

Figure 3 : ARXPS reconstructed composition profile of a $Al_2O_3/SiO_2/Si$ stack.

Next to layer thickness, it is also critical to determine the layer compositions and a composition profile. Only a few methods are available for a routine determination of composition profiles for very thin layers. Depth profiling methods such as (ToF)-SIMS can be easily applied but any sputter-based technique suffers from significant artefacts such as preferential sputtering effects. It is thus advised to rely on non-destructive methods. One of them is Medium Energy Ion Scattering

(MEIS) which is unfortunately not widely available. An interesting alternative can be found in ARXPS, which uses the strong dependence of signal intensities on material depth and, through an optimisation algorithm, allows to retrieve composition profiles from the angular difference in intensities of the different elements. An evaluation of this method has been documented in several publications[6,7,8].

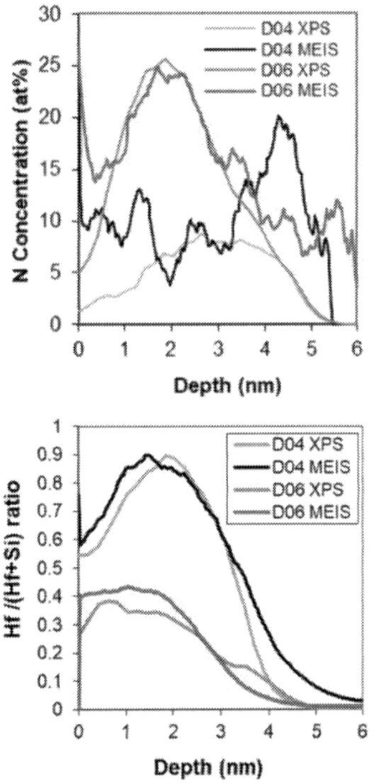

Figure 4 : ARXPS reconstructed composition profile of two nitrided HfSiO stacks.

For a simple layer structure such as the ~3.0 nm Al_2O_3/1.5 nm SiO_2/Si shown in Figure **3**, the composition profile is straightforward. However, obtaining reliable composition profiles in case of complex systems involves including some additional constraints such as stoichiometric ones. A typical complex system in gate stacks consists in the determination of composition profile of nitrided $Hf_xSi_{(1-x)}O_2$ layers deposited on Si. These types of samples have two important challenges as the Hf/(Hf+Si) may vary with depth and a peak overlap is present between the N 1s and an Hf loss peaks, making peak deconvolution challenging for low nitrogen concentration. Figure **4** presents the reconstructed profile for two stacks with different nominal Hf/Si ratios and nitridation processes. Sample D04 has a 3:1 Hf/Si ratio while sample D06 has a 1:3 ratio. Sample D04 is nitrided with a remote plasma process while sample D06 is nitrided using a NH_3 post-annealing process. The agreement between the Hf concentration profiles determined by ARXPS or MEIS is good. More discrepancies are observed in the Nitrogen concentration profile. The main differences are observed close to the surface and at the HfSiO/Si interface. At the interface, the main issue arises from the use of channelling during the MEIS experiment.

4 A second paradigm shift: Going 3D and full stack analysis

4.1 Analysing interfacial interactions

In the search of ultimate scaling, interfacial effects are critical, and it has been shown that the full high-k/metal gate stack needs to be considered as upon processing, the interface between the substrate and the high-k can be modified (for instance by oxygen scavenging) by the presence of the metal. Unfortunately, when the full stack is formed, the substrate/high-k interface may be beyond reach for regular photoemission measurements with Al Kα radiation[9]. At first, buried interfaces were studied using back-side XPS, i.e. by removing the substrate using polishing and etching technologies. This is however a difficult and very slow process. In addition, it is not applicable to all samples as an etch stop layer is needed under the stack to be analysed.

The alternative to back-side analysis is to use HAXPES, which with its high energy X-ray radiation allows the analysis of much deeper interfaces. HAXPES has long been limited to synchrotrons which severely limited its use for process development that requires frequent and easy access to the characterisation methods. Recently, several spectrometers were developed allowing high quality HAXPES using Ag Lα (2.98 keV)[10], Cr Kα (5.41 keV)[11] or Ga Kα (9.25 keV)[12] radiation.

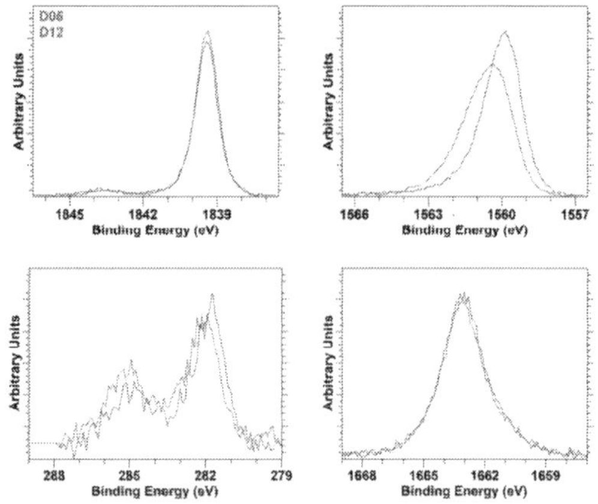

Figure 5 : Si1s (upper left), Al1s (upper right) C1s (lower left) and Hf $3d_{5/2}$ (lower right) HAXPES spectra of a ~12 nm thick TiN/TiAlC/TiN/HfO_2/SiO_2 stack on Si substrate, before (D05) and after (D12) annealing recorded with Cr Kα

In order to evidence the advantages of HAXPES, a multilayer sample with a total thickness of ~12 nm was processed and analysed before and after annealing. The analysed stack consists of 3 nm TiN/5 nm TiAlC/1 nm TiN/1.7 nm HfO_2/1 nm SiO_2/Si substrate and the corresponding Si 1s, Al 1s, C 1s, Hf $3d_{5/2}$ HAXPES curves before (D05) and after (D12) annealing are presented in Figure **5**. The Si and Hf spectra are not modified upon annealing showing a good chemical stability of the layers. However, significant changes are observed on the Al 1s spectra. A significant broadening and a shift towards high binding energy are observed, which is a clear indicator of chemical changes in the TiAlC layer.

4.2 3D-device technology

In present technologies such as FinFET, devices are developed in 3D and more complex structures are considered. As an example, Figure **6** shows a nanosheet structure with progressive lateral etching of the SiGe layers in repetitive Si/SiGe stacks. In order to avoid registering time consuming Transmission Electron Microscope (TEM) images for the monitoring of the SiGe etch, several top-down techniques were investigated, one of them being HAXPES using Cr Kα radiation. Thanks to the larger electron escape depth of HAXPES compared to XPS this type of analysis is within reach of PES. When monitoring Ge signal intensities, a good correlation between the measured Ge signal and the etch depth is observed, at the condition that the top SiO_2/SiN stack was first removed.

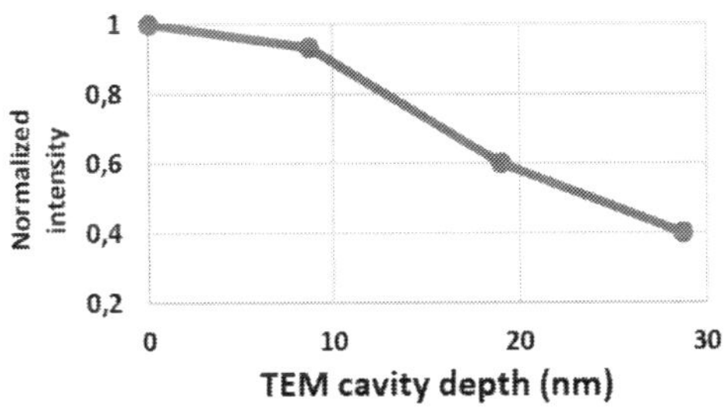

Figure 6 : Nanosheet structures (top) etched to various lateral depths and corresponding changes of the Ge 2p HAXPES curve intensities as measured with Cr Kα radiation

5 Future technologies

Future performance improvements along the path of Moore's law will imply further development in material integration and process capabilities.
For instance, the ability to pattern ever-smaller feature sizes and denser pitches is significantly challenged due to the lack of a fundamental understanding of EUV sub-picosecond exposure processes. Getting this understanding requires developments in analytical capabilities, including time resolved capabilities. Similar to the evolution of HAXPES from synchrotron to lab environment, time-resolved photoemission by use of a laser set-up will become readily available. Such a set-up is now available within imec and a more profound understanding on photoresist chemical modifications is developing through comparison of recorded spectra with theoretically modelled photoemission spectra. An example of modifications to experimental resist spectra upon photon exposure is shown in Figure 7.

These time-resolved capabilities will allow the characterization of the molecular and quantum dynamics of materials within the attosecond to picosecond for instance in 2D-materials or topological insulators.

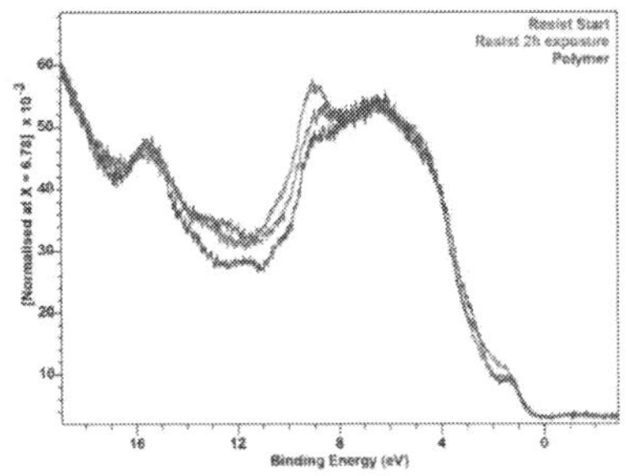

Figure 7 : UPS spectra from a non-exposed polymer-based resist (red), the base polymer (blue) and the exposed resist (green).

6 Conclusion

Photoemission is a very versatile characterisation technique that developed into a "must have" in the semiconductor industry. The technological evolution in instrumentation at lab-scale has led to a widening of the application field throughout the years. First, the introduction of parallel ARXPS allowed to easily retrieve depth information and composition depth profiling. Then, the introduction of high-quality HAXPES systems allowed to characterise buried interfaces in realistic systems as well as getting bulk-information. In the future, a laser-based system will allow time-resolved analysis, paving the road to a fundamental understanding of photoresist modifications and new materials.

[1] Howard R. Huff, Characterization and Metrology for ULSI Technology: 2000 International Conference, edited by D. G. Seiler, A. C. Diebold, T. J. Shaffner, R. McDonald, W. M. Bullis, P. J. Smith, and E. M. Secula, 2001 American Institute of Physics
[2] Moore, Gordon E. (1965-04-19). Electronics, Vol. 38, Number 8, April 19, 1965
[3] https://en.wikichip.org/wiki/technology_node
[4] P. Mack, R.G. White, J. Wolstenholme, T. Conard, Applied Surface Science 252 (2006) 8270–8276
[5] M. L. Green, M.-Y. Ho, B. Busch, G. D. Wilk, T. Sorsch, T. Conard, B. Brijs, W. Vandervorst, P. I. Räisänen, D. Muller, M. Bude, and J. Grazul, J. Appl. Phys., Vol. 92, No. 12, 15 December 2002, 7168
[6] T. Conard, W. Vandervorst, W (Vandervorst, W.) A. Bergmaier, K. Kimura, K; JVST A, Vol. 30; 31509, 2012
[7] K. Kimura, K. Nakajima, M. Zhao, H. Nohira, T. Hattori, M. Kobata, E. Ikenaga, J.J Kim, K. Kobayashi, T. Conard. and W. Vandervorst, Surf. Interface Anal. 2008; 40: 423–426
[8] A. Herrera-Gomez et al. Surf. Interface Anal.2009,41, 840–857
[9] Takashi Ando, Materials 2012, 5, 478-500
[10] https://www.kratos.com/products/axis-supra-xps-surface-analysis-instrument
[11] https://www.ulvac-phi.com/en/products/xps/quantes/
[12] https://scientaomicron.com/en/system-solutions/electron-spectroscopy/HAXPES-Lab

XPS chemical state mapping in opto- and microelectronics

Mathieu Frégnaux[1*], *Yoan* Bourlier[1,2], *Bruno* Berini[2], *Yves* Dumont[2], *Damien* Aureau[1]

[1]Institut Lavoisier de Versailles (ILV), CNRS UMR 8180, Université de Versailles Saint-Quentin-en-Yvelines - Université Paris-Saclay, 78000 Versailles, France.
[2]Groupe d'Etude de la Matière Condensée (GEMaC), CNRS UMR 8635, Université de Versailles Saint-Quentin-en-Yvelines - Université Paris-Saclay, 78000 Versailles, France.

Abstract. The strength of XPS imaging lies in its ability to (i) locate small patterns on sample surface, and (ii) inform, with micrometric lateral resolution, about the chemical environment of the elements detected at the surface. In this context, strontium-based perovskites appear to be well-adapted for such photoemission experiments thanks to their tunability and variability. These functional oxides have great potential for emerging opto- and microelectronic applications, especially for transparent conductive oxide. Patterned heterostructure $SrTiO_3/SrVO_3$ was grown by pulsed laser deposition using a shadow mask. This stack was then analysed by XPS mapping in serial acquisition mode. Ti2p and V2p core level imaging clearly highlights the $SrTiO_3$ and $SrVO_3$ domains. The XPS mapping of the Sr3d core level will be extensively discussed: strontium being a common element to both oxides with a very similar chemical environment. Despite a lower contrast in Sr3d images, the two materials are discernible thanks to the topography. In addtion, the use of Sr3d FWHM image is a real asset to evidence the two phases. Finally, data processing by principal component analysis allows us to extract significant spectral information on the strontium atoms.

1 Introduction

X-ray Photoelectron Spectroscopy (XPS) imaging is a real asset to access surface chemical compositions of a few millimeters square patterns. This specific photoemission mode can be useful for (i) confidently locating micrometric objects, or small features at the sample surface (ii) informing on the spatial distribution of the surface constituent elements (elemental mapping) and (iii) locally differentiating between chemical environments. There are two approaches for obtaining XPS images [1]: mapping (serial acquisition) or parallel imaging (parallel acquisition). These different acquisition modes were developed thanks to the reduction of X-ray spot sizes (down to few microns) or the joint use of a lens and a 2D detector, respectively.
Serial acquisition is based on a two-dimensional, rectangular array of small-area XPS [2,3] analyses. The lateral resolution is then determined by the size of the smallest analysis area available (usually few micrometers). Serial acquisition is generally slower compared to parallel acquisition. The data are collected using snapshot (acquired range of energies controlled by the pass energy) at each pixel while only one energy is recorded for parallel acquisition. The maximum size of the image is quite large and is only limited by the range of motion of the specimen stage
Parallel imaging exploits a different way for data acquisition. It images simultaneously the whole field of view for one fixed energy (fixed voltage applied). It requires additional lenses and a two-dimensional detector. The spatial resolution of parallel imaging depends on the spherical aberrations in the lens. Limiting the angular acceptance of the lens can reduce the effect of the aberrations, which improves resolution at the expense of sensitivity. The use of a magnetic immersion lens in the specimen region also reduces aberrations, producing higher sensitivity at a given resolution. Parallel Imaging provides the best resolution and is faster than serial methods for producing an image at a single energy. Spectroscopy from images is possible by collecting a series of images at different energies across the required energy range. A XPS spectrum can be then reconstructed for each pixel in the image.
An alternative approach is parallel imaging using X-ray photoelectron emission microscopy (XPEEM). The lateral resolution of core-level images in the 500 nm range is then achievable [4]. This is a real improvement compared with the capabilities of other lab XPS imaging methods. It uses a bright microfocussed monochromated Al Kα source and a photoelectron emission microscope (PEEM) as entrance lens of a high-transmission, aberration-compensated imaging spectrometer in the form of a double hemispherical analyser. From the instrumental point of view, the main limitation of the technique is the low counting statistics due to the use of a moderately bright millimeter size X-ray spot.
This reduction in lateral resolution makes XPS imaging compatible with the characterization of patterns in a wide range of industrial applications [1] and particularly in the field of micro-optoelectronics: SiGe layers [4], graphene [5], 2D materials [6] transparent conductive oxides [7]. The characteristic dimensions involved for

* Corresponding author: mathieu.fregnaux@uvsq.fr

such applications, especially thicknesses, require a probe size too small to be characterized with scanning electron microscopy coupled with energy dispersive X-ray spectroscopy (SEM-EDS) for instance. Moreover, the deposition of insulating passivation layers (oxides, nitrides, etc.) makes the use of Auger electron spectroscopy (AES) quite difficult to implement [8]. Under these conditions, despite a lower lateral resolution, XPS imaging with charge neutralization assistance turns out to be a technique of choice.

In contrast to the conventionally held belief of imaging XPS, data acquisition with good signal-to-noise ratio in a reasonable time is possible. Spectra from individual pixel of the image can be plotted or averaged for chemical information. Significant spectral information can be extracted from images, even with low count rate, using mathematical processes [9] such as principal component analysis (PCA).

In this paper, we will focus on XPS imaging in the context of perovskite oxide heterostructure characterization. This class of materials are promising for the emerging field of oxitronics [10] or the development of new transparent conductive oxide [11]. The advantages of serial acquisition that are its ability to produce high resolution, large area XPS images within minutes, will be exploited for this study. For this purpose, a patterned heterostructure $SrTiO_3/SrVO_3$ (STO/SVO) grown by pulsed laser deposition (PLD) on STO substrate will be investigated. Using a shadow mask, micrometer size domains of epitaxial STO with different shapes can be obtained on free SVO.

2 Experimental

SVO and STO targets for PLD were prepared in the GEMaC laboratory from high purity binary oxides powders (SrO/V_2O_5 and SrO/TiO_2) by standard ceramic processing method. The pressed pellet was sintered in air with additional oxygen flow. The "Sr/V" and "Sr/Ti" target metallic ratios were controlled by chemical titration. Single crystal (SurfaceNet GmbH) STO (100) substrates was used for the PLD growth. Depositions of both SVO and STO layers were performed in an ultra-high vacuum chamber with base pressure of 5×10-7 Pa. A shadow mask was applied during the STO deposition to create patterns of various shape and size ranging from millimetres to few ten microns. Targets were ablated using a KrF laser with a wavelength of 248 nm and 20 ns of pulse duration. The laser fluency, the focalization lens position and the laser energy were previously optimized for individual materials. The targeted thicknesses were 20 nm for SVO and 80 nm for STO.

Regular X-ray Photoelectron Spectroscopy (XPS) measurements were carried out using a Thermofisher Scientific Nexsa spectrometer with a monochromatic Al-Kα X-ray source (hν = 1486.6 eV). The detection was performed perpendicularly to the sample surface using a constant analyzer energy (CAE) mode (pass energy 20 eV) and spectra were recorded with a 0.1 eV energy step. The use of low-energy electron and ion flood gun was necessary to perform the analysis. Imaging of the STO/SVO stack was recorded in the "SnapMap" mode with a 50 μm X-ray spot. In such acquisition mode, the micro-focused beam is rastered across the sample by rapid stage movements, yielding an elementary map (O1s, Sr3d, Ti2p and V2p) with a field of view of 3 x 3 mm². Indeed, the stage movement is synchronized with the spectrometer which collects snapshot XPS spectra continuously during rastering. Full spectra are acquired at each pixel with pass energy of 150 eV. Each map required 14 minutes of acquisition time (4 scans). In order to quantify the concentrations of the elements across the images (atomic concentration images), the spectra extracted in each pixel were used to define quantification regions Quantification and image processing including PCA were performed using the Thermofisher Scientific Avantage© software. Chemical compositions and quantitative chemical maps were obtained from the O1s, Sr3d, Ti2p, V2p peak areas after a Shirley type background subtraction and considering "AlThermo1" sensitivity factor library. Concerning PCA data processing, two factors were defined and used. No data scaling was employed but a offset was applied to remove a horizontal baseline at the minimum value of data in each spectrum. Non-negativity fitting was imposed in order to constrain the rotation of factors to produce positive factors.

3 Results and discussion

Fig.1 shows Sr3d spectral regions of both pure STO (black line) and SVO (red line) thin films deposited by PLD on STO substrate. These reference spectra were recorded by regular XPS. In the case of STO substrates, the Sr3d doublet of this reference material is well-defined. The two peaks $Sr3d_{5/2}$ and $Sr3d_{3/2}$ are narrow and separated by 1.76 eV spin orbit splitting. The main peak is centred at 133.4 eV with a full width at half maximum (FWHM) of 0.8 eV. On the contrary, pure SVO thin film exhibits a broad doublet (1.7 eV FWHM) shifted to higher binding energy compared to STO.

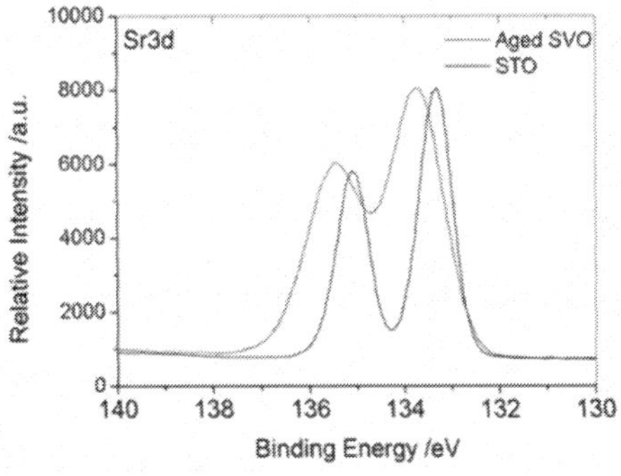

Fig. 1. Sr3d spectral regions of both STO (black) and air-exposed SVO (red) recorded by regular XPS on PLD thin films

It is important to note that this XPS characterization was performed without particular precaution prior analysis. The sample was exposed to air, not store under vacuum or inert atmosphere As SVO surfaces are known to be highly reactive [11,12] and very sensitive to external stimuli [13,14], the obtained Sr3d spectrum in Fig. 1 is the sum of several contributions such as SVO effective signature but also Sr-rich phases, hydroxides or carbonates. This preliminary work will be essential for the characterization of the STO/SVO heterostructure by serial mapping. The main challenge here will rely in processing the data from the Sr3d core level mapping. Indeed, strontium is common to both oxides with very similar chemical environments, *i.e.* very close in binding energy (see Fig. 1). A robust fitting procedure should be required to extract chemical state mapping of strontium. Fig.2 shows XPS area maps (3 mm x 3 mm) of the STO/SVO stack for Ti2p (a) and V2p (b) spectral regions. The quantitative core level imaging highlights both STO and SVO domains since Ti and V are discriminating elements. The chemical contrast is sharp. This means that no signal from underneath layers is detected. STO patterns cover well the SVO thin film.

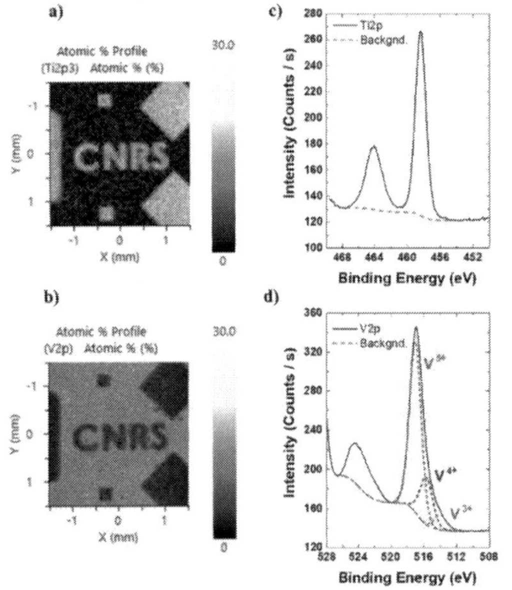

Fig. 2. Quantitative XPS maps (3 mm x 3 mm) of the STO/SVO stack for Ti2p (a) and V2p (b) elements and their corresponding integrated spectrum (c and d). $V2p_{3/2}$ peak is reconstructed to evidence vanadium oxidation states.

Extracted spectrum from Ti2p map (Fig.2 c) is a doublet with $Ti2p_{3/2}$ component at 458.4 eV which is attributed to Ti^{4+}, as expected in STO. Concerning V2p map, the corresponding integrated spectrum (Fig. 2 d) is also a doublet. It is dominated by an intense component at 516.9 eV which corresponds to V^{5+} oxidation state. Small amounts of V^{4+} and V^{3+} are detected at lower binding energy (1.7 and 3.4 eV below) [15]. However, to achieve electrical neutrality in SVO phase, vanadium is expected to be V^{4+}. This discrepancy can be explained by the heterostructure elaboration process. SVO growth was followed by the deposition of STO (under a low oxygen partial pressure). Since vanadium tends to oxidize easily, a small layer with a higher oxidation state V^{5+} on top of vanadate material (likely V_2O_5 phases) has grown during STO deposition. In these conditions, strontium-rich nanostructures can even appear at the surface of SVO. Previous studies [12] demonstrate that vanadium atoms have a high oxidation state (V^{5+}) in these nanostructures.

Fig. 3 a shows quantitative XPS Sr3d map (3 mm x 3 mm) on the same region of interest of the STO/SVO heterostructure. Patterns are no longer discernible since strontium is common to both oxides in similar proportions (20% at.). However, a topographic contrast is visible on the area map (Fig. 3 c) since the STO patterns are deposited on top of the SVO layer providing 80 nm height difference. The corresponding integrated spectrum is presented in Fig. 3 b. It exhibits a broad doublet with a maximum centred at 133.6 eV. Unfortunately, it does not allow us to easily distinguish the respective contributions of STO and SVO.

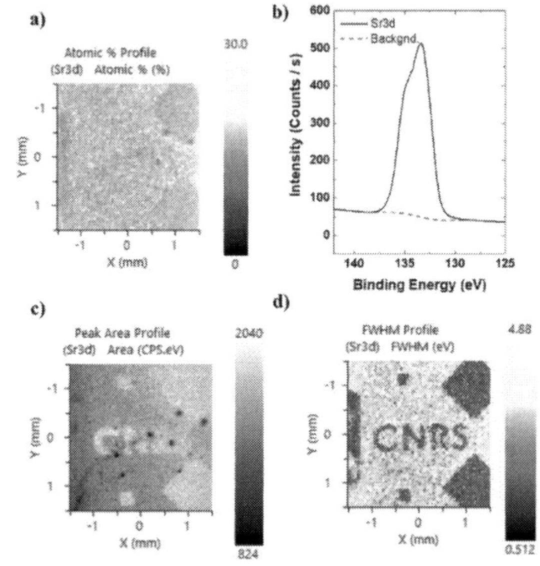

Fig. 3. Quantitative XPS map (3 mm x 3 mm) of the STO/SVO stack for Sr3d (a) and its corresponding integrated spectrum (b). Sr3d area (d) and FWHM (d) maps on the same region of interest.

As demonstrated in reference material (Fig. 1), the positions of Sr3d peak in STO and SVO are close but they present different FWHM. In order to take advantage of this spectroscopic feature, Fig. 3 d presents FWHM map of the STO/SVO stack for Sr3d spectral region. The reasonably good contrast obtained clearly evidences domains with low FWHM associated to STO and others with higher FWHM values attributed to SVO. To extract more information of the Sr3d core level image and get additional knowledge on the chemical environment of the probed strontium atoms in both oxides, a PCA mathematical process was applied. The results obtained are shown in Fig. 4 a-d.

First, two eigenvectors arbitrary named PCA 1 and PCA 2 were sufficient to reconstruct the Sr3d image. They are plotted in Fig. 4 a. Even if these vectors are pure mathematical objects, their shapes satisfactory look like Sr3d doublet. PCA 1 and PCA 2 have a maximum intensity at 133.4 and 132.5 eV, respectively. These positions are really close to the ones measured for $Sr3d_{5/2}$ in STO and SVO, respectively.

Overlayed PCA profiles are presented in Fig 3 (b) while individual PCA 1 and PCA 2 profiles are show in Fig 3 (c) and (d). STO patterns are perfectly reconstructed by PCA 1 profile. The image is similar to the one obtained on Ti2p core level (Fig 2. a), STO discriminating element. Indeed, Sr3d spectrum recorded in regular XPS on STO patterns (Fig. 4 d) is comparable to the eigenvector PCA 1 (fit with one doublet at 133.4 eV).

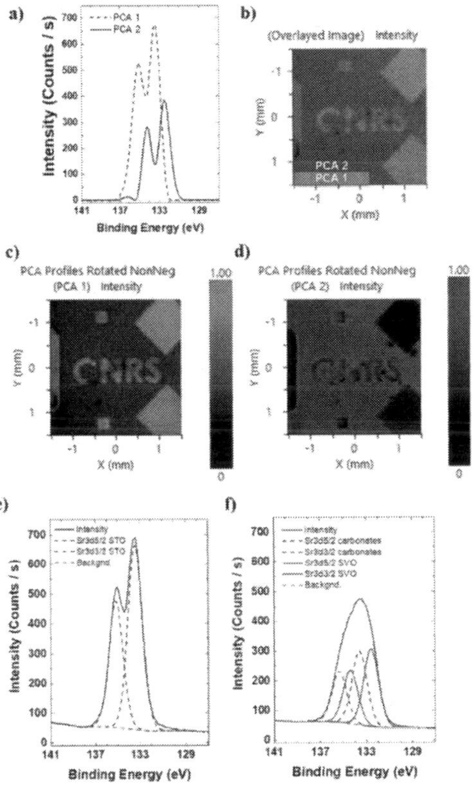

Fig. 4. Sr3d image treatment using PCA mathematical process: (a) eigenvectors considered (PCA 1 and PCA 2), (b) overlayed PCA profiles (c) PCA 1 profile, (d) PCA 2 profile, Sr3d spectrum obtained by regular XPS in STO patterns (e) and free SVO (f)

At first sight, PCA 2 profile would reconstruct the regions with free SVO but it is not the case. A linear combination of 60% PCA 2 with 40% PCA 1 must be considered. This treatment has similarities with the data obtained by regular XPS on free SVO area (Fig. 4 e). Indeed, two doublets are needed to obtain SVO signature: the one at lower binding energy (132.6 eV) is attributed to effective conductive SVO phase while the one at higher binding energy (133.6 eV) corresponds to strontium hydroxydes or strontium carbonates [12].

4 Conclusion

Serial acquisition was used to produce high resolution, large area (3 mm x 3 mm) XPS images of a patterned heterostructure STO/SVO grown by PLD on STO substrate. Using a shadow mask, micrometer size domains of epitaxial STO with different shapes were obtained on free SVO. The Ti2p and V2p area imaging clearly highlights both STO and SVO domains since titanium and vanadium are discriminating elements. The chemical contrast is sharp. This means meaning that no signal from underneath layers is detected and STO patterns entirely cover well the free SVO thin film. Spectroscopy from images confirms the expected oxidation state Ti^{4+} for the titanium in STO and the multiple valence of vanadium dominated by V^{5+} for air-exposed SVO surface. The XPS chemical state mapping of the Sr3d core level is challenging since strontium is a common element to both oxides with a very similar chemical environment. Despite lower contrast in Sr3d images, the two materials are discernible thanks to topography. As the FWHM of air exposed SVO is rather higher than STO one, Sr3d FWHM image is a simple way to evidence the two phases. Moreover, PCA mathematical process was applied on the Sr3d core level image. Two eigenvectors were sufficient to reconstruct the Sr3d image. One of them is enough to reconstruct STO patterns whereas free SVO needs the linear combination of the two eigenvectors. This mathematical processing shows similarities with spectroscopic data.

Acknowledgements:
The authors thank the French Agence Nationale de la Recherche (ANR) in the framework of the POLYNASH project (ANR-17-CE08-0012) for Dr Yoan Bourlier postdoctoral position funding.

References

1. D. J. Morgan, Journal of Electron Spectroscopy and Related Phenomena **231**, 109 (2019).
2. K. Yates and R. H. West, Surface and Interface Analysis **5**, 217 (1983).
3. C. Demanet, Vacuum **37**, 465 (1987).
4. O. Renault, M. Lavayssière, A.Bailly, D. Mariolle et al.., Journal of Electron Spectroscopy and Related Phenomena **171,** 68–71 (2009).
5. A. Tyagi et al., Nanoscale, **14**, 2167-2176 (2022).
6. M. Frégnaux H. Kim, D. Rouchon et al., Surface and Interface Analysis **48**, 465–469 (2016).
7. H. Piao, L. Le Tarte, W.A. Hennessy N. Fairley, Surface and Interface Analysis, **3,** 493-500 (2007).
8. J. Wolstenholme, *Auger Electron Spectroscopy: Practical Application to Materials Analysis and Characterization of Surfaces, Interfaces, and Thin Films* (Momentum Press, 2015).
9. J. Walton and N. Fairley, *The Casa Cookbook Part 2: XPS Image Processing* (Acolyte Science, Knutsford, Cheshire, UK, 2011).
10. M. Coll et al. Applied Surface Science **482,** 1 (2019).
11. A. Boileau, A. Cheikh, A. Fouchet, A. David, C. Labbé, P. Marie, F. Gourbilleau, and U. Lüders, Advanced Optical Materials **7**, 1801516 (2019).
12. Y. Bourlier, M. Frégnaux, B. Bérini, A. Fouchet, Y. Dumont, and D. Aureau, Applied Surface Science **553**, 149536 (2021).
13. Y. Bourlier, M. Frégnaux, B. Bérini, A. Fouchet, Y. Dumont, and D. Aureau, ChemNanoMat **5**, 674 (2019).
14. K. Ridier, D. Aureau, B. Bérini, Y. Dumont, N. Keller, J. Vigneron, A. Etcheberry, B. Domengès, and A. Fouchet, Phys. Rev. B **97**, 035146 (2018).
15. C. Lin, A. Posadas, T. Hadamek, and A. A. Demkov, Phys. Rev. B **92**, 035110 (2015).

High resolution and time resolved photoemission spectroscopy for developing more efficient materials to reduce energy consumption and increase renewable energy production.

Mathieu G. Silly[1*]

[1]Synchrotron SOLEIL, L'Orme des Merisiers Départementale 128 91190 Saint-Aubin, France

Abstract. Due to the increase of energy consumption and the resulting ecological challenge, a collective awareness leads to the development of renewable energies and more efficient materials to increase the green energy production. Development of efficient photovoltaic materials is very closely related to their chemical and electronic properties. A better knowledge of these imbricated properties is needed, in addition to a better comprehension of their interplay with charge transport mechanisms. Exciton creation and recombination processes, charge transfer and charge collection processes take place at the surface and interface of the photoactive materials. Photoemission spectroscopy as chemical specific and surface sensitive spectroscopic technique is a method of choice on the study of physical phenomena at the origin of photoconversion efficiency. Time resolved photoemission spectroscopy has been recently renewed interest covering time scale from fs to more than seconds. It permits to probe the dynamics of relaxation of photoexcited charges and determine their lifetime. It finds application in various materials used in solar photovoltaics. In this paper, we define the physical and chemical properties determined by the combination of high resolution and time resolved photoemission spectroscopy. We show examples dealing with the development of renewable energy and energy consumption reduction in agreement with the current ecological trend for a better future.

1 Introduction

The primary energy consumption has increased by 5.8% in 2021 and fossil fuels accounts for 83% of the global primary consumption.[1] Use of fossil energy leads to irrevocable phenomena as temperature increase which could led to dramatic events. International awareness led to ecological chart (Paris agreement of the *Paris climate* conference (*COP21*)) which actively participates in the development of renewable energy and in energy consumption reduction. The share of renewables continuously increases to represent 13% in global power generation.[1] To develop, renewable energy must be economically competitive. Energy production cost reduction is needed. Due to the huge effort done the last 10 years, the photovoltaics energy cost has been divided by a factor of four, by reducing the material cost, by improving the efficiency and reliability.[2]

Promising new materials such as colloidal nanoparticles[3] and perovskite[4, 5] have been developed. As charge creation and collection processes take place at the surface and interfaces of materials, core level photoemission spectroscopy is a powerful tool due to its chemical selectivity and surface sensitivity. It is possible to determine the chemistry of the surfaces and interfaces. The location of the different chemical species at the interface can be assessed by varying the kinetic energy of the probed photoelectron. This is allowed by synchrotron radiation sources and their capability to tune the incident photon energy. In the early 2000s, time resolved photoemission spectroscopy (TR-PES) has been used to probe the charge dynamics in semiconductors through surface photovoltage measurement.[6, 7] With third generation synchrotron centres, new beamlines combining high resolution photoemission and time resolved photoemission spectroscopy offer new capabilities appealing for the development of photoactive materials.[8-10] Based on new delay line detectors[11] high resolution photoelectron analysers[12] combined to high resolution, soft X-ray beamline offers resolving power better than 15000 and time resolutions down to 10ps.[13] The covered temporal range benefits of exploring charge carrier recombination dynamics, charge diffusion and thermal diffusion.[14] New setups based on HHG and XFEL offer better time resolutions in the fs range to probe carrier excitation and thermalization processes.[14] In this paper we will focus on the capabilities offered by synchrotron-based experiments taking advantages of high resolution and time resolution photoemission spectroscopy. We will show different results on bulk materials with use in photovoltaic applications, hybrid materials as organic-inorganic heterojunctions, colloidal nanoparticles, and their applications.

* Corresponding author: mathieu.silly@synchrotron-soleil.fr

2 Photoemission process

Basic concepts of photoemission spectroscopy will be described, more details can be found in literature.[15, 16] Photoemission spectroscopy is based on photoelectric effect. The surface of the sample is irradiated with X-rays. Photoelectrons are ejected from the sample with a kinetic energy verifying the equation [1] in agreement with the law of conservation of energy.

$$h\nu = E_B + E_K + \phi \qquad [1]$$

Where $h\nu$ is the incident photon energy, E_B is the binding energy, E_K is the kinetic energy and ϕ is the work function of the material. The photoelectrons will be collected through high resolution photoelectron analyser and sorted according to their kinetic energy. The photoelectrons will stem from three main processes. An elastic signal made of sharp peaks whom energy varies with the incident photon energy originates from given atomic orbitals. Broader peaks with kinetic energy independent on incident photon energy come from Auger deexcitation processes. A background originates from inelastic processes. From the elastic process, chemical and electronic properties can be determined that helps provide a better understanding and optimisation of material for renewable energy.

2.1 Core level photoemission spectroscopy

Core level photoemission spectroscopy is focused on the collection of the photoelectrons coming from given atomic orbital. The photoelectrons are collected as a function of their kinetic energy. The number of contributions needed to fit given core level spectrum give the number of non-equivalent chemical species in the material. The position in energy of the peaks (chemical shift) provides information on the nature of the chemical bond and of the surrounding atoms of the species. The spatial origin of the collected photoelectron into the bulk can also be determined by changing the kinetic energy of the photoelectrons. Photoelectrons have an inelastic means free path varying as a function of their kinetic energy and the nature of the materials crossed. The inelastic means free path can be calculated using Tanuma Powell and Penn algorithm.[17]. About 95% of the collected photoelectrons are coming from 3 times the electron means free path. By varying the incident photon energy, it is possible to vary the kinetic energy of the photoelectrons and carry out a depth profiling. By comparing the variation in relative intensities of the core levels as a function of the incident photon energy, it is possible to determine the nature, number and amount of inequivalent chemical species in the material as well as their location at the interface. From the variation in intensity of the C1s core level spectra measured as a function of the incident photon energy, the intercalation process of silicon at graphene/SiC interface has been successfully determined.[18]

2.2 Valence band maximum and cut off

The determination of the band diagram energy is of great importance for efficient charge carrier creation and collection. It allows to determine the type of band alignment for semiconductor heterojunction. The position in energy of the valence and conduction bands between two materials involved in active P-N or donor-acceptor junctions provides information on the ability to dissociate the photocharge carriers. From the measurement of the valence band, the valence band maximum energy position is determined. The work function of the material is determined from the photoemission signal Cut Off corresponding to the photoelectrons measured at lowest kinetic energy.[19] It is also possible to evidence the formation of interfacial dipole between two materials, by comparing the evolution in energy position of the work function and of the core levels before and after deposit of a second material on top of a first one (atoms, organic molecules, 2D materials on semiconductor or metal).[20] Interfacial dipole formation can drastically affect the electronic properties of the formed heterojunction.

2.3 Surface photovoltage effect

The surface photovoltage (SPV) effect is a powerful contactless and non-destructive technique to determine the minority charge carrier lifetime.[21] The measurement of SPV in photoemission is based on band bending phenomenon.[22] Band bending comes from the difference in charge between the surface and the bulk. The position of the Fermi level into the gap of semiconductors, in the bulk, is directly related to the doping level of the material. The position of the Fermi level at the surface depends on the surface termination. At the equilibrium, an electric field is created in the depletion layer (DL). Under light excitation with an energy higher than the gap of the material, charges are promoted from the valence band to the conduction band. Due to the electric field in the DL, the electrons are attracted or repelled from the surface to compensate the excess of charges and flattened the bands. The spectroscopic signature of this phenomenon is a rigid shift of the core levels. The direction of the energy shift depends on the substrate doping. The doping of a material can be determined from the direction of the shift observed under illumination. The time needed to come back to the steady state position after illumination gives the lifetime of the photoexcited charges.

3 Materials for photovoltaic and LED applications

3.1 Bulk semiconductors

3.1.1 Silicon

Silicon is a material of choice in photovoltaic application and represents 90% of photovoltaic market.[23] Silicon solar cells has reached an energy conversion efficiency of 25% in 1999 and new architectures have been developed to exceed this value. It has been shown that minority charge carrier lifetime plays a key role in photoconversion efficiency. A minority charge carrier lifetime optimum exists.[24] Nevertheless, the parameters driving the minority charge carrier lifetimes remain unclear. Experimental works paved the way to formalisms[6, 7], but no systematic studies were done to determine the influence of the key parameters governing photocarriers lifetime. Taking advantages of new time resolved photoemission setup at synchrotron, systematic studies on surface termination and substrate doping have been performed. Their influence on SPV effect has been determined and theoretical models have been evaluated to fit the experimental data.

High resolution and time resolved photoemission spectroscopy has been performed on In/Si(111) surface as a function of indium coverage and surface reconstruction.[25] Depending on the surface termination, a broad dispersion in energy of the position of the Si2p core level has been obtained corresponding to a variation in band bending. The SPV relaxation exhibits two deexcitation lifetimes. A fast deexcitation process (≤100ns) corresponds to a tunnelling transport process. The photoexcited charges cross the surface potential barrier through tunnel effect. The size of the depletion layer increases, the tunnel recombination is then no more allowed. A second longer deexcitation process occurs based on thermionic effect (≥100ns). This works evidenced the key role played by the size of the depletion layer in the relaxation process and relaxation lifetime.

The influence of silicon substrate doping and surface termination on SPV has also been systematically studied by time resolved photoemission spectroscopy.[26] Water saturated and thermally oxidized Si(001) surfaces have been studied as a function of bulk doping. It has been shown that bulk doping has the highest influence on charge carrier lifetime. Thermionic relaxation model is in good agreement with experimental data for low laser fluence and bulk recombination becomes preponderant at high laser fluence as proposed in literature.[27] Experimental SPV shift as a function of the laser fluence agrees with current theoretical model at low fluence but differs at high fluence. New model is needed to account for SPV shift saturation observed at high fluence.

3.1.2 wide band gap semiconductors

ZnO as n-type wide bang gap semiconductor (3.4eV) with high electron mobility[28] is a promising material as transparent electrodes or as active layer for solar cell application.[29] It nevertheless suffers from persistent photoconductivity. Its origin is attributed to oxygen vacancies.[30]

SPV effect in ZnO has been investigated by time resolved photoemission spectroscopy (TR-PES).[22] The implication of oxygen vacancies in SPV effect in ZnO has been evidenced by changing the surface preparation procedure. The annealing of the ZnO surface in partial pressure of oxygen for different times has shown dramatic changes in the relaxation lifetime of SPV. Annealing of 20 min and 10 min in 1.2-1.4x10^{-7} mbars of oxygen at 703 K led to relaxation lifetime of 150 µs and 1.2 ms, respectively. For sub bandgap excitation of ZnO at 405 nm (3.06 eV), a small SPV shift is still observed. Even if the laser pulse energy is too small to photoexcite the charges through the band gap, ZnO presents SPV effect. These results indicate that the photoexcitation process involves in-gap states associated to ionized oxygen vacancies. This work has corroborated the involvement of ionized oxygen vacancies with persistent photoconductivity in ZnO.

3.2 Heterostructures

The combination of different materials is of great interest to build efficient photoactive heterostructures made of donor-acceptor junction for organic-organic junction[31] or p-n junctions for inorganic heterojunction. Due to their high tunability, it is possible to match the electronic properties of organic materials with the electronic properties of inorganic semiconductor. The formation of organic-inorganic heterostructure is a promising approach. Inorganic material as ZnO can act as active layer or only play part of transparent electrode.[29] A key process in charge generation and collection is the charge transfer occurring at the interface between two materials.

3.2.1 hybrid heterostructures

Model donor-acceptor organic junction made of ZnPc/C60 multilayers has been investigated by TR-PES to probe the charge dynamics at the ps time scale.[32] To be sensitive to the molecular orbitals of the molecules, valence band measurement has been performed at the carbon absorption edge. Working at the resonance, signal coming from the molecular orbital in the valence band is enhanced. A special care on the cleanness of the optics of the soft X-ray beamlines is needed to allow photoemission measurement at Carbon K edge with high photon flux.[33, 34] Time resolved photoemission spectroscopy measured with time resolution of 50 ps as a function of laser flux has evidenced two photoexcitation processes. A rigid shift of the VB at high laser flux coming from space charge effect and a variation in intensity of the valence band at high energy attributed to a modification of the interface dipole.

Charge relaxation in C60/CuPc system appears complex. Different studies using TR-PES evidenced various behavior depending on the stacking of C60/CuPc and also on the nature of the substrate underneath. TR-PES measured on C60 on thin CuPc film evidenced localized photoexcited charges on C60 layer.[35] C1s core level contribution coming from the C60 was shifted under laser excitation when the contribution coming from CuPc remained unchanged. This result evidenced the capability to TR-PES to probe charge transfer localized on specific chemical species. The time relaxation of photoexcited charges has been investigated on C60/CuPc/TiO2 and CuPc/C60/TiO2 heterojunctions.[36] The experiments have evidenced different relaxation behaviors for the two systems. A delay in exciton dissociation is observed for C60/CuPc/TiO2 compared to CuPc/C60/TiO2. This indicated the role of charge blocking of the CuPc layer disabling the exciton relaxation through TiO2. This work has highlighted the influence of molecular stacking order in organic junction on relaxation process.

3.2.2 colloidal quantum dots

Colloidal Quantum Dots (CQDs) present advantageous characteristics as tuneable electronic properties (bandgap, doping,) and low-cost material production.[3, 37] CQDs find application in solar cell application as active material but also application in the energy consumption reduction in efficient structure for LED applications[38]. CQDs take advantages of high resolution and time resolved photoemission spectroscopy to optimise the production process and material tuning to adapt the electronic properties and to match electronic properties of their counterparts (P-N junctions).

Carrying out depth profiling by measuring the evolution of the relative intensity of core levels as a function of incident photon energy, it is possible to have a qualitative idea of the repartition of chemical species in the nanoparticles. The synthesis of InP/ZnS core shell nanoparticles has been investigated by high resolution core level photoemission spectroscopy (HR-PES).[39] By comparing the spectroscopic signature for two synthesis methods, it has been possible to determine the chemical species repartition inside the nanoparticles. While a single-step synthesis method has led to a homogenous repartition of Indium species forming InPZnS alloyed core nanoparticles, the two-step method has led to the synthesis of InP/ZnS core shell structure. Photoemission spectroscopy has also evidenced interfacial states induced by species interdiffusion at core/shell interface.

It is also possible to determine quantitatively the thickness of the shell according to the core diameter by HR-PES.[40] Using analytical model developed by Shard et al.[41, 42] considering the spherical geometry of the nanoparticles, by plotting the variation in intensity between two chemical species coming from the core and the shell and by comparing that to the calculation, it is possible to determine the thickness of the shell. HR-PES is also very efficient in the optimisation of ligand exchange process. Ligand exchange process permits to exchange long Carbon chain used during the CQDs synthesis with shorter conductive ligands used to dope the nanoparticles. Anchor is usually of different chemical nature. For instance, in the case of PbS/CdS CQDs,[40] long carbon chains were linked through nitrogen atoms while shorter ligands were linked through Sulfuric atoms. HR-PES spectra have evidenced two kinds of nitrogen atoms and two kinds of sulfuric atoms corresponding to ligands linked to the nanoparticles and free ligands present into the media. By comparing the ratio of the different chemical species, it is possible to determine the amount of remaining ligands and to examine the ligand exchange process efficiency.[40]

The doping of CQDs can be performed by ligand exchange or by changing the CQDs stoichiometry. The doping of the CQDs can be determined by measuring the VBM and by determining the energy band diagram. The doping can also be directly determined from the direction of SPV shift under laser excitation. Both ligand exchange process and chemical composition modification take advantages of HR- and TR- PES to evidence doping variation in CQDs. n to p doping of HgTe CQDs by ligand exchange process has been evidenced by combination of HR-PES and TR-PES.[43] Continuous variation in HgSexTe1-x nanoparticles doping from n to p type by varying the nanoparticle composition has been evidenced by energy band diagram determination and TR-PES measurement.[44]

Development of CQDs based solar cells strongly depends on charge collection efficiency. Determining the CQDs/electrode electronic coupling is of high interest. To evidence indirect substrate excitation via charge transfer from the CQDs to the substrate, PbS QDs absorbing in red light have been deposited on clean ZnO substrate.[45] An illumination in red light of PbS/ZnO surface produced a shift in energy of the PbS core level comparable to an excitation in UV light above the ZnO band gap. No shift in energy was visible for red illumination of ZnO clean surface. These results indicated that QDs photoexcitation allows for charge transfer into the ZnO conduction band at the origin of transient core level energy shift. TR-PES allowed to evidence charge transfer from QDs to ZnO surface and indirect ZnO excitation under illumination.

4 Conclusion and perspectives

To increase the photoconversion efficiency and go beyond the Shockley–Queisser limit[46], corresponding to the maximum theoretical efficiency of a single p-n junction solar cell, new materials and new architectures are developed.[3] High resolution and time resolved photoemission spectroscopy has proved its capability to determine the chemical and electronic properties of complex nanomaterials and their chemical and electronic coupling with the substrate underneath. Synchrotron based photoemission spectroscopy combining variable incident photon energy, high energy resolution and time resolution down to 10 ps, is a powerful and unique spectroscopic technique able to correlate chemical and electronic properties to charge carrier lifetime. New technological advances in TR-PES experimental setup arise. Higher time resolution at fs time scale is now available and allows to probe faster excitation and fast decay processes as thermalization effects (carrier-carrier and carrier-phonon scattering)[14]. Time resolved near ambient pressure environment[9] is also developed with use in photocatalysis and photochemistry. Both will participate to develop more efficient materials to decrease the energy consumption and especially in the scope of catalysis as catalytic processes account for 25% of industrial energy use.[47]

References

(1) bp. *Statistical Review of World Energy, 71st edition*; 2022. https://www.bp.com/content/dam/bp/business-sites/en/global/corporate/pdfs/energy-economics/statistical-review/bp-stats-review-2022-full-report.pdf.

(2) Office of Energy Efficiency & Renewable Energy, https://www.energy.gov/eere/solar/photovoltaics (accessed.

(3) Carey, G. H.; Abdelhady, A. L.; Ning, Z.; Thon, S. M.; Bakr, O. M.; Sargent, E. H. Colloidal Quantum Dot Solar Cells. *Chemical Reviews* **2015**, *115* (23), 12732-12763. DOI: 10.1021/acs.chemrev.5b00063.

(4) Jena, A. K.; Kulkarni, A.; Miyasaka, T. Halide Perovskite Photovoltaics: Background, Status, and Future Prospects. *Chemical Reviews* **2019**, *119* (5), 3036-3103. DOI: 10.1021/acs.chemrev.8b00539.

(5) Abd Mutalib, M.; Ludin, N. A.; Ruzalman, N. A. A. N.; Barrioz, V.; Sepeai, S.; Teridi, M. A. M.; Su'ait, M. S.; Ibrahim, M. A.; Sopian, K. Progress towards highly stable and lead-free perovskite solar cells. *Materials for Renewable and Sustainable Energy* **2018**, *7* (2). DOI: 10.1007/s40243-018-0113-0.

(6) Widdra, W.; Brocker, D.; Giessel, T.; Hertel, I. V.; Kruger, W.; Liero, A.; Noack, F.; Petrov, V.; Pop, D.; Schmidt, P. M.; et al. Time-resolved core level photoemission: surface photovoltage dynamics of the SiO2/Si(100) interface. *Surface Science* **2003**, *543* (1-3), 87-94, Article. DOI: 10.1016/j.susc.2003.07.005.

(7) Brocker, D.; Giessel, T.; Widdra, W. Charge carrier dynamics at the SiO2/Si(100) surface: a time-resolved photoemission study with combined laser and synchrotron radiation. *Chemical Physics* **2004**, *299* (2-3), 247-251, Article. DOI: 10.1016/j.chemphys.2003.11.028.

(8) Polack, F.; Silly, M.; Chauvet, C.; Lagarde, B.; Bergeard, N.; Izquierdo, M.; Chubar, O.; Krizmancic, D.; Ribbens, M.; Duval, J. P.; et al. TEMPO: a New Insertion Device Beam line at SOLEIL for Time Resolved Photoelectron Spectroscopy Experiments on Solids and Interfaces. In *10th International Conference on Synchrotron Radiation Instrumentation*, Melbourne, AUSTRALIA, Sep 27-Oct 02, *2010*; *AIP Conference Proceedings*: Vol. 1234, pp 185-188. DOI: 10.1063/1.3463169.

(9) Shavorskiy, A.; Neppl, S.; Slaughter, D. S.; Cryan, J. P.; Siefermann, K. R.; Weise, F.; Lin, M. F.; Bacellar, C.; Ziemkiewicz, M. P.; Zegkinoglou, I.; et al. Sub-nanosecond time-resolved ambient-pressure X-ray photoelectron spectroscopy setup for pulsed and constant wave X-ray light sources. *Review of Scientific Instruments* **2014**, *85* (9), 8, Article. DOI: 10.1063/1.4894208.

(10) Ogawa, M.; Yamamoto, S.; Kousa, Y.; Nakamura, F.; Yukawa, R.; Fukushima, A.; Harasawa, A.; Kondoh, H.; Tanaka, Y.; Kakizaki, A.; et al. Development of soft x-ray time-resolved photoemission spectroscopy system with a two-dimensional angle-resolved time-of-flight analyzer at SPring-8 BL07LSU. *Review of Scientific Instruments* **2012**, *83* (2), 7, Article. DOI: 10.1063/1.3687428.

(11) Cautero, G.; Sergo, R.; Stebel, L.; Lacovig, P.; Pittana, P.; Predonzani, M.; Carrato, S. A two-dimensional detector for pump-and-probe and time resolved experiments. *Nuclear Instruments & Methods in Physics Research Section a-Accelerators Spectrometers Detectors and Associated Equipment* **2008**, *595* (2), 447-459, Article. DOI: 10.1016/j.nima.2008.06.046.

(12) Bergeard, N.; Silly, M. G.; Krizmancic, D.; Chauvet, C.; Guzzo, M.; Ricaud, J. P.; Izquierdo, M.; Stebel, L.; Pittana, P.; Sergo, R.; et al. Time-resolved photoelectron spectroscopy using synchrotron radiation time structure. *Journal of Synchrotron Radiation* **2011**, *18*, 245-250. DOI: 10.1107/s0909049510052301.

(13) Silly, M. G.; Ferte, T.; Tordeux, M. A.; Pierucci, D.; Beaulieu, N.; Chauvet, C.; Pressacco, F.; Sirotti, F.; Popescu, H.; Lopez-Flores, V.; et al. Pump-probe experiments at the TEMPO beamline using the low-alpha operation mode of Synchrotron SOLEIL. *Journal of Synchrotron Radiation* **2017**, *24*, 886-897. DOI: 10.1107/s1600577517007913.

(14) Sundaram, S. K.; Mazur, E. Inducing and probing non-thermal transitions in semiconductors using femtosecond laser pulses. *Nature Materials* **2002**, *1* (4), 217-224. DOI: 10.1038/nmat767.

(15) Stevie, F. A.; Donley, C. L. Introduction to x-ray photoelectron spectroscopy. *Journal of Vacuum Science & Technology A* **2020**, *38* (6). DOI: 10.1116/6.0000412.

(16) Ratner, B. D.; Castner, D. G. Electron Spectroscopy for Chemical Analysis. In *Surface Analysis – The Principal Techniques*, 2nd ed.; John C. Vickerman , I. S. G. Ed.; John Wiley & Sons, Ltd, 2009; pp 47-112.

(17) Tanuma, S.; Powell, C. J.; Penn, D. R. Calculations of electron inelastic mean free paths. V. Data for 14 organic compounds over the 50–2000 eV range. *Surf. Interface Anal.* **1994**, *21*, 165–176. DOI: 10.1002/sia.740210302.

(18) Silly, M. G.; D'Angelo, M.; Besson, A.; Dappe, Y. J.; Kubsky, S.; Li, G.; Nicolas, F.; Pierucci, D.; Thomasset, M. Electronic and structural properties of graphene-based metal-semiconducting heterostructures engineered by silicon intercalation. *Carbon* **2014**, *76*, 27-39. DOI: 10.1016/j.carbon.2014.04.033.

(19) Gao, Y. L. Surface analytical studies of interfaces in organic semiconductor devices. *Materials Science & Engineering R-Reports* **2010**, *68* (3), 39-87. DOI: 10.1016/j.mser.2010.01.001.

(20) Pierucci, D.; Mahmoudi, A.; Silly, M.; Bisti, F.; Oehler, F.; Patriarche, G.; Bonell, F.; Marty, A.; Vergnaud, C.; Jamet, M.; et al. Evidence for highly p-type doping and type II band alignment in large scale monolayer WSe2/Se-terminated GaAs heterojunction grown by molecular beam epitaxy. *Nanoscale* **2022**, *14* (15), 5859-5868. DOI: 10.1039/d2nr00458e.

(21) Kronik, L.; Shapira, Y. Surface photovoltage phenomena: theory, experiment, and applications. *Surface Science Reports* **1999**, *37* (1-5), 1-206, Review. DOI: 10.1016/s0167-5729(99)00002-3.

(22) Spencer, B. F.; Graham, D. M.; Hardman, S. J. O.; Seddon, E. A.; Cliffe, M. J.; Syres, K. L.; Thomas, A. G.; Stubbs, S. K.; Sirotti, F.; Silly, M. G.; et al. Time-resolved surface photovoltage measurements at n-type photovoltaic surfaces: Si(111) and ZnO(10(1)over-

bar0). *Physical Review B* **2013**, *88* (19), 16, Article. DOI: 10.1103/PhysRevB.88.195301.
(23) Andreani, L. C.; Bozzola, A.; Kowalczewski, P.; Liscidini, M.; Redorici, L. Silicon solar cells: toward the efficiency limits. *Advances in Physics-X* **2019**, *4* (1). DOI: 10.1080/23746149.2018.1548305.
(24) Jayakrishnan, R.; Gandhi, S.; Suratkar, P. Correlation between solar cell efficiency and minority carrier lifetime for batch processed multicrystalline Si wafers. *Materials Science in Semiconductor Processing* **2011**, *14* (3-4), 223-228. DOI: 10.1016/j.mssp.2011.02.020.
(25) Ogawa, M.; Yamamoto, S.; Fujikawa, K.; Hobara, R.; Yukawa, R.; Kitagawa, S.; Pierucci, D.; Silly, M. G.; Lin, C. H.; Liu, R. Y.; et al. Relaxations of the surface photovoltage effect on the atomically controlled semiconductor surfaces studied by time-resolved photoemission spectroscopy. *Physical Review B* **2013**, *88* (16), 9, Article. DOI: 10.1103/PhysRevB.88.165313.
(26) Pierucci, D.; Silly, M. G.; Tissot, H.; Hollander, P.; Sirotti, F.; Rochet, F. Surface photovoltage dynamics at passivated silicon surfaces: influence of substrate doping and surface termination. *Faraday Discussions* **2022**. DOI: 10.1039/d1fd00107h.
(27) Long, J. P.; Sadeghi, H. R.; Rife, J. C.; Kabler, M. N. SURFACE SPACE-CHARGE DYNAMICS AND SURFACE RECOMBINATION ON SILICON(111) SURFACES MEASURED WITH COMBINED LASER AND SYNCHROTRON RADIATION. *Physical Review Letters* **1990**, *64* (10), 1158-1161. DOI: 10.1103/PhysRevLett.64.1158.
(28) Kaidashev, E. M.; Lorenz, M.; von Wenckstern, H.; Rahm, A.; Semmelhack, H. C.; Han, K. H.; Benndorf, G.; Bundesmann, C.; Hochmuth, H.; Grundmann, M. High electron mobility of epitaxial ZnO thin films on c-plane sapphire grown by multistep pulsed-laser deposition. *Applied Physics Letters* **2003**, *82* (22), 3901-3903. DOI: 10.1063/1.1578694.
(29) Vittal, R.; Ho, K.-C. Zinc oxide based dye-sensitized solar cells: A review. *Renewable & Sustainable Energy Reviews* **2017**, *70*, 920-935. DOI: 10.1016/j.rser.2016.11.273.
(30) Lany, S.; Zunger, A. Anion vacancies as a source of persistent photoconductivity in II-VI and chalcopyrite semiconductors. *Physical Review B* **2005**, *72* (3). DOI: 10.1103/PhysRevB.72.035215.
(31) Scharber, M. C.; Sariciftci, N. S. Efficiency of bulk-heterojunction organic solar cells. *Progress in Polymer Science* **2013**, *38* (12), 1929-1940. DOI: 10.1016/j.progpolymsci.2013.05.001.
(32) Costantini, R.; Pincelli, T.; Cossaro, A.; Verdini, A.; Goldoni, A.; Cichon, S.; Caputo, M.; Pedio, M.; Panaccione, G.; Silly, M. G.; et al. Time resolved resonant photoemission study of energy level alignment at donor/acceptor interfaces. *Chemical Physics Letters* **2017**, *683*, 135-139. DOI: 10.1016/j.cplett.2017.04.033.
(33) Chauvet, C.; Polack, F.; Silly, M. G.; Lagarde, B.; Thomasset, M.; Kubsky, S.; Duval, J. P.; Risterucci, P.; Pilette, B.; Yao, I.; et al. Carbon contamination of soft X-ray beamlines: dramatic anti-reflection coating effects observed in the 1 keV photon energy region. *Journal of Synchrotron Radiation* **2011**, *18*, 761-764. DOI: 10.1107/s0909049511023119.
(34) Yao-Leclerc, I.; Brochet, S.; Chauvet, C.; De Oliveira, N.; Duval, J. P.; Gil, J. F.; Kubsky, S.; Lagarde, B.; Nahon, L.; Nicolas, F.; et al. Handling the carbon contamination issue at SOLEIL. In *Damage to Vuv, Euv, and X-Ray Optics Iii*, Juha, L., Bajt, S., London, R. A. Eds.; Proceedings of SPIE, Vol. 8077; 2011.
(35) Arion, T.; Neppl, S.; Roth, F.; Shavorskiy, A.; Bluhm, H.; Hussain, Z.; Gessner, O.; Eberhardt, W. Site-specific probing of charge transfer dynamics in organic photovoltaics. *Applied Physics Letters* **2015**, *106* (12), 5, Article. DOI: 10.1063/1.4916278.
(36) Ozawa, K.; Yamamoto, S.; Mase, K.; Matsuda, I. A Surface Science Approach to Unveiling the TiO2 Photocatalytic Mechanism: Correlation between Photocatalytic Activity and Carrier Lifetime. *E-Journal of Surface Science and Nanotechnology* **2019**, *17*, 130-147. DOI: 10.1380/ejssnt.2019.130.
(37) Sargent, E. H. Colloidal quantum dot solar cells. *Nature Photonics* **2012**, *6* (3), 133-135. DOI: 10.1038/nphoton.2012.33.
(38) Bossavit, E.; Qu, J.; Abadie, C.; Dabard, C.; Dang, T.; Izquierdo, E.; Khalili, A.; Greboval, C.; Chu, A.; Pierini, S.; et al. Optimized Infrared LED and Its Use in an All-HgTe Nanocrystal-Based Active Imaging Setup. *Advanced Optical Materials* **2022**, *10* (4). DOI: 10.1002/adom.202101755.
(39) Huang, K.; Demadrille, R.; Silly, M. G.; Sirotti, F.; Reiss, P.; Renault, O. Internal Structure of InP/ZnS Nanocrystals Unraveled by High-Resolution Soft X-ray Photoelectron Spectroscopy. *Acs Nano* **2010**, *4* (8), 4799-4805. DOI: 10.1021/nn100581t.
(40) Clark, P. C. J.; Radtke, H.; Pengpad, A.; Williamson, A. I.; Spencer, B. F.; Hardman, S. J. O.; Leontiadou, M. A.; Neo, D. C. J.; Fairclough, S. M.; Watt, A. A. R.; et al. The passivating effect of cadmium in PbS/CdS colloidal quantum dots probed by nm-scale depth profiling. *Nanoscale* **2017**, *9* (18), 6056-6067. DOI: 10.1039/c7nr00672a.
(41) Shard, A. G.; Wang, J.; Spencer, S. J. XPS topofactors: determining overlayer thickness on particles and fibres. *Surface and Interface Analysis* **2009**, *41* (7), 541-548. DOI: 10.1002/sia.3044.
(42) Shard, A. G. A Straightforward Method For Interpreting XPS Data From Core-Shell Nanoparticles. *Journal of Physical Chemistry C* **2012**, *116* (31), 16806-16813. DOI: 10.1021/jp305267d.
(43) Livache, C.; Izquierdo, E.; Martinez, B.; Dufour, M.; Pierucci, D.; Keuleyan, S.; Cruguel, H.; Becerra, L.; Fave, J. L.; Aubin, H.; et al. Charge Dynamics and Optolectronic Properties in HgTe Colloidal Quantum Wells. *Nano Letters* **2017**, *17* (7), 4067-4074. DOI: 10.1021/acs.nanolett.7b00683.
(44) Dabard, C.; Planelles, J.; Po, H.; Izquierdo, E.; Makke, L.; Greboval, C.; Moghaddam, N.; Khalili, A.; Dang, T. H.; Chu, A.; et al. Optimized Cation Exchange for Mercury Chalcogenide 2D Nanoplatelets and Its Application for Alloys. *Chemistry of Materials* **2021**, *33* (23), 9252-9261. DOI: 10.1021/acs.chemmater.1c02951.
(45) Spencer, B. F.; Leontiadou, M. A.; Clark, P. C. J.; Williamson, A. I.; Silly, M. G.; Sirotti, F.; Fairclough, S. M.; Tsang, S. C. E.; Neo, D. C. J.; Assender, H. E.; et al. Charge dynamics at heterojunctions for PbS/ZnO

colloidal quantum dot solar cells probed with time-resolved surface photovoltage spectroscopy. *Applied Physics Letters* **2016**, *108* (9). DOI: 10.1063/1.4943077.
(46) Shockley, W.; Queisser, H. J. DETAILED BALANCE LIMIT OF EFFICIENCY OF P-N JUNCTION SOLAR CELLS. *Journal of Applied Physics* **1961**, *32* (3), 510-519. DOI: 10.1063/1.1736034.
(47) Friend, C. M.; Xu, B. Heterogeneous Catalysis: A Central Science for a Sustainable Future. *Accounts of Chemical Research* **2017**, *50* (3), 517-521. DOI: 10.1021/acs.accounts.6b00510.

AUTHOR INDEX

EDP Sciences – Web of Conferences
17, Avenue du Hoggar
Parc d'Activité de Courtabouf
BP 112
F-91944 Les Ulis Cedex A
France

ISBN 978-1-7138-6621-3

CURRAN ASSOCIATES INC.
proceedings.com